ENERGY SCIENCE, ENGINEERING AND TECHNOLOGY

NUCLEAR ENERGY

RESEARCH, DEVELOPMENT AND NEW TECHNOLOGIES ROADMAP

ENERGY SCIENCE, ENGINEERING AND TECHNOLOGY

Additional books in this series can be found on Nova's website under the Series tab.

Additional E-books in this series can be found on Nova's website under the E-book tab.

ENERGY SCIENCE, ENGINEERING AND TECHNOLOGY

NUCLEAR ENERGY

RESEARCH, DEVELOPMENT AND NEW TECHNOLOGIES ROADMAP

PETR MAKRYGINA
EDITOR

Nova Science Publishers, Inc.
New York

For permission to use material from this book please contact us:
Telephone 631-231-7269; Fax 631-231-8175
Web Site: http://www.novapublishers.com

Library of Congress Cataloging-in-Publication Data

Nuclear energy : research, development, and new technologies roadmap / editor, Petr Makrygina.
p. cm.
Includes index.
ISBN 978-1-61470-999-2 (hardcover)
1. Nuclear energy--United States. 2. Nuclear power plants--United States--Design and construction. 3. Nuclear power plants--United States--Safety measures. I. Makrygina, Petr.
TK9023.N79 2011 621.48'30973--dc23 2011030127

Published by Nova Science Publishers, Inc. † New York

CONTENTS

PREFACE

This book provides a roadmap for the Department of Energy's (DOE's) Office of Nuclear Energy (NE) research, development and demonstration activities that will ensure nuclear energy remains a viable energy option for the United States.

Chapter 1- To achieve energy security and greenhouse gas (GHG) emission reduction objectives, the United States must develop and deploy clean, affordable, domestic energy sources as quickly as possible. Nuclear power will continue to be a key component of a portfolio of technologies that meets our energy goals. This document provides a roadmap for the Department of Energy's (DOE's) Office of Nuclear Energy (NE) research, development, and demonstration activities that will ensure nuclear energy remains viable energy option for the United States.

Chapter 2- The Nuclear Energy Enabling Technologies (NEET) Program proposed for fiscal year (FY) 2011 at the requested \$99 million will develop crosscutting technologies that directly support and complement the Department of Energy, Office of Nuclear Energy's (DOE–NE's) advanced reactor and fuel cycle concepts, focusing on innovative research that offers the promise of dramatically improved performance. Pending FY 2011 Congressional Appropriation, the program will encourage the development of transformative, "out-of-the-box" solutions across the full range of nuclear energy technology, spurring revolutionary improvements in safety, performance, reliability, economics, and proliferation risk reduction.

In: Nuclear Energy
Editor: Petr Makrygina

ISBN: 978-1-61470-999-2

Chapter 1

NUCLEAR ENERGY RESEARCH AND DEVELOPMENT ROADMAP*

United States Department of Energy

LIST OF ACRONYMS

BTU	British Thermal Units
CO_2	Carbon dioxide
DOE	Department of Energy
EE	DOE–Office of Energy Efficiency and Renewable Energy
EIA	Energy Information Agency
EPRI	Electric Power Research Institute
FE	DOE–Office of Fossil Energy
GDP	Gross domestic product
GHG	Greenhouse gas
GWe	Gigawatt (electric)
GWe-yr	Gigawatt-year (electric)
HTGR	High-temperature gas-cooled reactor
HTR	High-temperature reactor
IAEA	International Atomic Energy Agency
II&C	Instrumentation, information and control

* This is an edited, reformatted and augmented version of a United States Department of Energy publication, dated April 2010.

IPSR	Integral primary system reactor
ITAAC	Inspections, test, analyses and acceptance criteria
kW-hr	Kilowatt-hour
LWR	Light-water reactor
MPACT	Materials Protection, Accounting and Control for Transmutation
MT	Metric ton
MWe	Megawatt (electric)
MWh	Megawatt-hour
NDE	Nondestructive evaluation
NE	DOE–Office of Nuclear Energy
NEA	Nuclear Energy Agency
NGNP	Next Generation Nuclear Plant
NGSI	Next Generation Safeguards Initiative
NNSA	National Nuclear Security Administration
NRC	Nuclear Regulatory Commission
OECD	Organization for Economic Cooperation and Development
R&D	Research and development
RISMC	Risk-informed safety margin characterization
SC	DOE–Office of Science
SMR	Small, modular reactor
UNF	Used nuclear fuel

EXECUTIVE SUMMARY

To achieve energy security and greenhouse gas (GHG) emission reduction objectives, the United States must develop and deploy clean, affordable, domestic energy sources as quickly as possible. Nuclear power will continue to be a key component of a portfolio of technologies that meets our energy goals. This document provides a roadmap for the Department of Energy's (DOE's) Office of Nuclear Energy (NE) research, development, and demonstration activities that will ensure nuclear energy remains viable energy option for the United States.

Today, the key challenges to the increased use of nuclear energy, both domestically and internationally, include:

- The capital cost of new large plants is high and can challenge the ability of electric utilities to deploy new nuclear power plants.

- The exemplary safety performance of the U.S. nuclear industry over the past thirty years must be maintained by an expanding reactor fleet.
- There is currently no integrated and permanent solution to high-level nuclear waste management.
- International expansion of the use of nuclear energy raises concerns about the proliferation of nuclear weapons stemming from potential access to special nuclear materials and technologies.

In some cases, there is a necessary and appropriate federal role in overcoming these challenges, consistent with the primary mission of NE to advance nuclear power as a resource capable of making major contributions to meeting the nation's energy supply, environmental, and energy security needs. This is accomplished by resolving technical, cost, safety, security and proliferation resistance barriers, through research, development, and demonstration, as appropriate. NE's research and development (R&D) activities will help address challenges and thereby enable the deployment of new reactor technologies that will support the current fleet of reactors and facilitate the construction of new ones.

Research and Development Objectives

NE organizes its R&D activities along four main R&D objectives that address challenges to expanding the use of nuclear power: (1) develop technologies and other solutions that can improve the reliability, sustain the safety, and extend the life of current reactors; (2) develop improvements in the affordability of new reactors to enable nuclear energy to help meet the Administration's energy security and climate change goals; (3) develop sustainable nuclear fuel cycles; and (4) understanding and minimization of risks of nuclear proliferation and terrorism.

R&D Objective 1. Develop Technologies and Other Solutions that can Improve the Reliability, Sustain the Safety, and Extend the Life of Current Reactors

The existing U.S. nuclear fleet has a remarkable safety and performance record, and today these reactors account for 70 percent of the low greenhouse gas (GHG)-emitting domestic electricity production. Extending the operating lifetimes of current plants beyond sixty years and, where possible, making further improvements in their productivity will generate near-term benefits. Industry has a significant financial incentive to extend the life of existing plants, and as such, activities will be cost shared. Federal R&D investments are appropriate to answer fundamental scientific questions and, where private investment is insufficient, to help make progress on broadly applicable technology issues that can generate public benefits. The DOE role in this R&D objective is to work in conjunction with industry and where appropriate the Nuclear Regulatory Commission (NRC) to support and conduct the long-term research needed to inform major component refurbishment and replacement strategies, performance enhancements, plant license extensions, and age-related regulatory oversight decisions. DOE will focus on aging phenomena and issues that require long-term research and are generic to reactor type.

R&D Objective 2. Develop Improvements in the Affordability of New Reactors to Enable Nuclear Energy to Help Meet the Administration's Energy Security and Climate Change Goals

If nuclear energy is to be a strong component of the nation's future energy portfolio, barriers to the deployment of new nuclear plants must be overcome. Impediments to new plant deployment, even for those designs based on familiar light-water reactor (LWR) technology, include the substantial capital cost of new plants and the uncertainties in the time required to license and construct those plants. Although subject to their own barriers for deployment, more advanced plant designs, such as small modular reactors (SMRs) and high-temperature reactors (HTRs), have characteristics that could make them more desirable than today's technology. SMRs, for example, have the potential to achieve lower proliferation risks and more simplified construction than other designs. The development of next-generation reactors could present lower capital costs and improved efficiencies. These reactors may be based upon new designs that take advantage of the advances in high performance computing while leveraging capabilities afforded by improved structural materials. Industry plays a substantial role in overcoming the barriers in this area. DOE provides support through R&D ranging from fundamental nuclear phenomena to the development of advanced fuels that could improve the economic and safety performance of these advanced reactors. Nuclear power can reduce GHG emissions from electricity production and possibly in co-generation by displacing fossil fuels in the generation of process heat for applications including refining and the production of fertilizers and other chemical products.

R&D Objective 3. Develop Sustainable Nuclear Fuel Cycles

Sustainable fuel cycle options are those that improve uranium resource utilization, maximize energy generation, minimize waste generation, improve safety, and limit proliferation risk. The key challenge is to develop a suite of options that will enable future decision makers to make informed choices about how best to manage the used fuel from reactors. The Administration has established the Blue Ribbon Commission on America's Nuclear Future to inform this waste- management decision-making process. DOE will conduct R&D in this area to investigate technical challenges involved with three potential strategies for used fuel management:

- *Once-Through* – Develop fuels for use in reactors that would increase the efficient use of uranium resources and reduce the amount of used fuel requiring direct disposal for each megawatt-hour (MWh) of electricity produced. Additionally, evaluate the inclusion of non-uranium materials (*e.g.*, thorium) as reactor fuel options that may reduce the long-lived radiotoxic elements in the used fuel that would go into a repository.
- *Modified Open Cycle* – Investigate fuel forms and reactors that would increase fuel resource utilization and reduce the quantity of long-lived radiotoxic elements in the used fuel to be disposed (per MWh), with limited separations steps using technologies that substantially lower proliferation risk.
- *Full Recycling* – Develop techniques that will enable the long-lived actinide elements to be repeatedly recycled rather than disposed. The ultimate goal is to develop a cost-effective and low proliferation risk approach that would dramatically decrease the long-term danger posed by the waste, reducing uncertainties associated with its disposal.

DOE will work to develop the best approaches within each of these tracks to inform waste management strategies and decision making.

R&D Objective 4. Understand and Minimize the Risks of Nuclear Proliferation and Terrorism

It is important to assure that the benefits of nuclear power can be obtained in a manner that limits nuclear proliferation and security risks. These risks include the related but distinctly separate possibilities that nations may attempt to use nuclear technologies in pursuit of a nuclear weapon and that terrorists might seek to steal material that could be used in a nuclear explosive device. Addressing these concerns requires an integrated approach that incorporates the simultaneous development of nuclear technologies, including safeguards and security technologies and systems, and the maintenance and strengthening of non-proliferation frameworks and protocols. Technological advances can only provide part of an effective response to proliferation risks, as institutional measures such as export controls and safeguards are also essential to addressing proliferation concerns. These activities must be informed by robust assessments developed for understanding, limiting, and managing the risks of nation-state proliferation and physical security for nuclear technologies. NE will focus on assessments required to inform domestic fuel cycle technology and system option development. These analyses would complement those assessments performed by the National Nuclear Security Administration (NNSA) to evaluate nation state proliferation and the international nonproliferation regime. NE will work with other organizations including the NNSA, the Department of State, the NRC, and others in further defining, implementing and executing this integrated approach.

R&D Areas

The Department expects to undertake R&D in a variety of areas to support its role in the objectives outlined above. Examples include:

- Structural materials
- Nuclear fuels
- Reactor systems
- Instrumentation and controls
- Power conversion systems
- Process heat transport systems
- Dry heat rejection
- Separations processes

- Waste forms
- Risk assessment methods
- Computational modeling and simulation

R&D Approach

A goal-driven, science-based approach is essential to achieving the stated objectives while exploring new technologies and seeking transformational advances. This science-based approach, depicted in Figure 1, combines theory, experimentation, and high-performance modeling and simulation to develop the fundamental understanding that will lead to new technologies. Advanced modeling and simulation tools will be used in conjunction with smaller-scale, phenomenon-specific experiments informed by theory to reduce the need for large, expensive integrated experiments. Insights gained by advanced modeling and simulation can lead to new theoretical understanding and, in turn, can improve models and experimental design. This R&D must be informed by the basic research capabilities in the DOE Office of Science (SC).

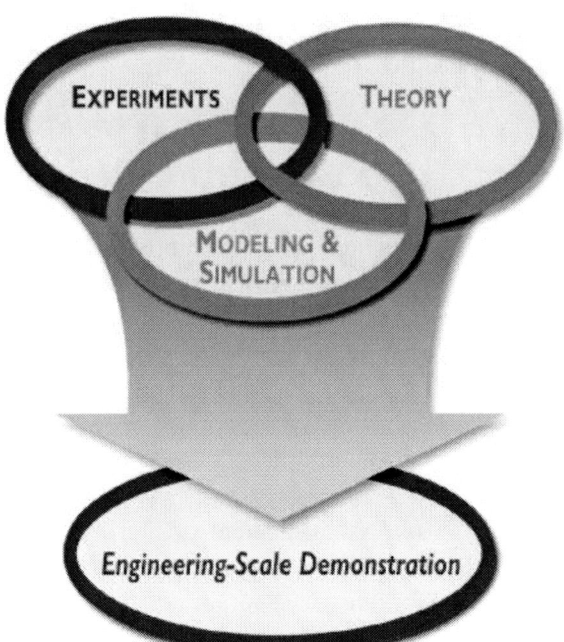

Figure 1. Major Elements of a Science-Based Approach.

NE maintains access to a broad range of facilities to support its research activities. Hot cells and test reactors are at the top of the hierarchy, followed by smaller-scale radiological facilities, specialty engineering facilities, and small non-radiological laboratories. NE employs a multi- pronged approach to having these capabilities available when needed. The core capabilities rely on DOE-owned irradiation, examination, chemical processing and waste form development facilities. These are supplemented by university capabilities ranging from research reactors to materials science laboratories. In the course of conducting this science-based R&D, infrastructure needs will be evaluated and considered through the established planning and budget development processes.

There is potential to leverage and amplify effective U.S. R&D through collaboration with other nations via multilateral and bilateral agreements, including the Generation IV International Forum. DOE is also a participant in Organization of Economic Cooperation and Development/Nuclear Energy Agency (OECD/NEA) and International Atomic Energy Agency (IAEA) initiatives that bear directly on the development and deployment of new reactor systems. In addition to these R&D activities, international interaction supported by NE and other government agencies will be essential in establishment of international norms and control regimes to address and mitigate proliferation concerns.

1. Introduction

Access to affordable, abundant energy – chiefly from fossil fuel sources – has been a key enabler of economic growth since the Industrial Revolution. However, as the first decade of the 21st century draws to a close, the United States finds itself confronted with economic, environmental, and national security challenges related in part to the manner in which our society produces, distributes, and uses energy. Continued access to plentiful, secure, and environmentally benign energy is fundamental to overcoming these challenges.

Nuclear energy is an important element of the diverse energy portfolio required to accomplish our national objectives. NE conducts research and development, and demonstrations, as appropriate, that will help enable the benefits of clean, safe, secure and affordable nuclear energy to continue and expand.

> Nuclear power is a proven clean, affordable, domestic energy source that is part of the current U.S. energy portfolio.

This document identifies opportunities and challenges associated with continued and increased use of fission energy to enhance our nation's prosperity, security, and environmental quality; outlines the NE role and mission in enabling the benefits of nuclear energy for our nation; and presents a strategy and roadmap to guide the NE scientific and technical agenda. The report presents a high-level vision and framework for R&D activities needed to keep the nuclear energy option viable in the near term and to expand its use in the decades ahead.

Section 2 describes the current energy production and utilization landscape in the United States. Section 3 articulates NE's fundamental mission and role in enabling nuclear energy solutions and presents the four R&D objectives for nuclear energy development that are the focus of NE activities. The details of the roadmap are presented in Section 4. The R&D approach presented in Section 5 embodies a goal-oriented, science-based R&D portfolio that includes both evolutionary and transformational, high-risk–high-payoff R&D, including those research areas that encompass multiple objectives. Finally, Section 6 provides a summary of the objects presented in this report.

This report is not an implementation plan, but rather provides a basis that will guide NE's internal programmatic and strategic planning for research going forward.

The report focuses on R&D activities sponsored by NE. The U.S. nuclear industry plays a central role in overcoming barriers and is ultimately responsible for the commercial deployment of the resulting technologies. NE intends to proceed in a manner that supports a strong and viable nuclear industry in the United States and preserves the ability of that industry to participate in nuclear projects here and abroad.

Finally, it should be noted that in some limited cases, NE's mission extends beyond terrestrial deployment of nuclear energy into other arenas, such as space applications of both fission and radioisotope power systems. Some technology development needs identified in this document also benefit space applications, but these mission arenas are not addressed in this roadmap. Educational programs, while vital, are interwoven through the technical programs and are not discussed as separate entities.

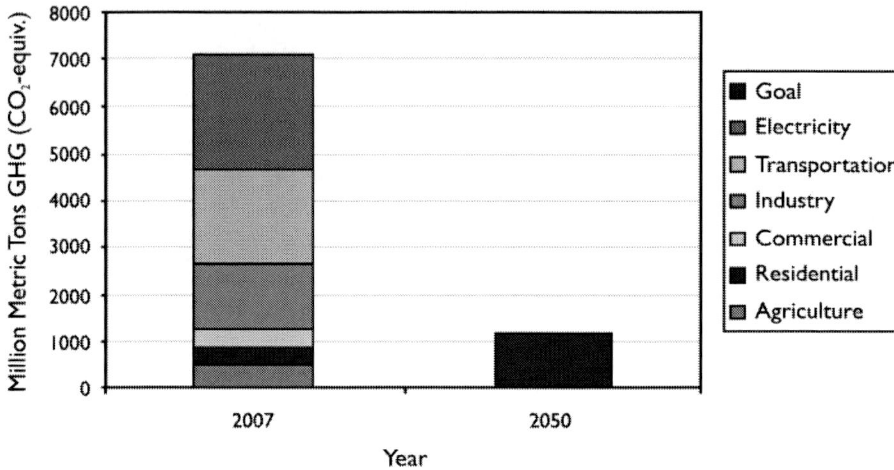

Figure 2. U.S. Greenhouse Gas Emissions.[1]

> To achieve its energy security and GHG reduction objectives, the U.S. must develop and deploy clean, affordable, domestic energy sources as quickly as possible.

2. Background

All governments of the world share a common challenge to ensure their people have access to affordable, abundant, and environmentally friendly energy. Secretary of Energy Steven Chu has reiterated the Administration's position that nuclear is an important part of the energy mix. He has recognized the importance of nuclear energy in meeting this challenge and supports R&D that can help increase the benefits of nuclear energy. A key objective that will shape the energy landscape of the United States is the transition to clean energy sources with reductions in GHG emissions (with a quantitative goal of 83% reduction below 2005 emissions levels by 2050, shown in Figure 2).

2.1. The Energy Landscape

The Human Development Index[2] is a commonly used measure of quality of life. Figure 3 illustrates that a nation's standard of living depends in part on energy consumption. Access to adequate energy is now and will continue to be required to achieve a high quality of life. Economic development, combined with efforts to limit carbon emissions, will likely lead to a significant expansion of nuclear power. The U.S., in concert with the international community, must develop the technologies and systems to accomplish such expansion while limiting proliferation risks.

As we move forward, efficiency and conservation will become ever-increasing components of energy policy. However, conservation and energy efficiency alone will not be sufficient to maintain a desirable quality of life.

The United States currently consumes roughly 100 quadrillion British Thermal Units (BTU), or 100 quads, of primary energy.[3] This represents 25% of world's energy consumption in a country that produces 30% of the global gross domestic product (GDP). Figure 4 shows energy consumption in the United States as a function of sectors and energy sources. At present, 40% of the total energy consumed is in the form of electricity, of which about 20 percent is generated by nuclear power. With 6 billion metric tons (MT) of emitted carbon dioxide (CO_2) as a result of fossil fuel usage (see Figure 5), the United States contributes 25 percent of global GHGs emitted.

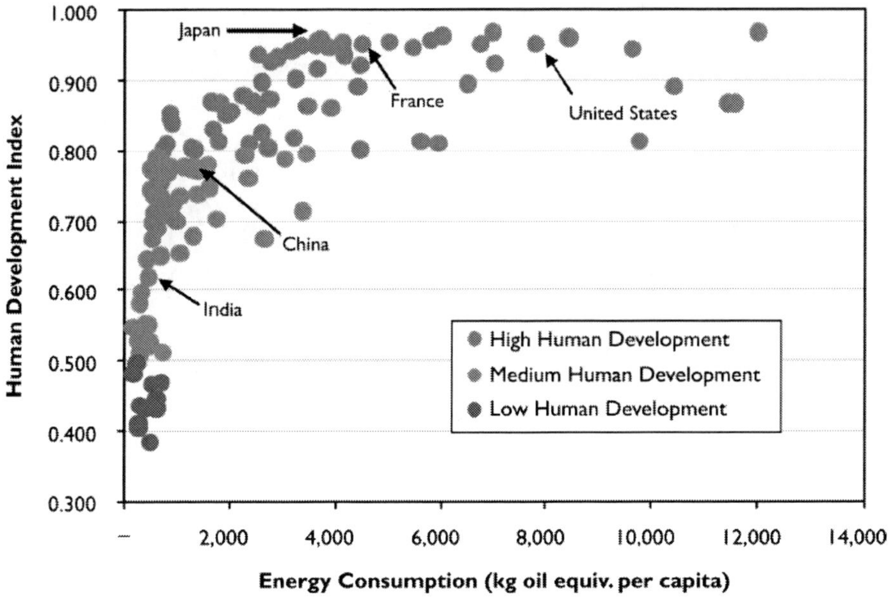

Figure 3. 2005 Human Development Index vs. Energy Consumption (Per Capita Kilograms Oil Equivalent).

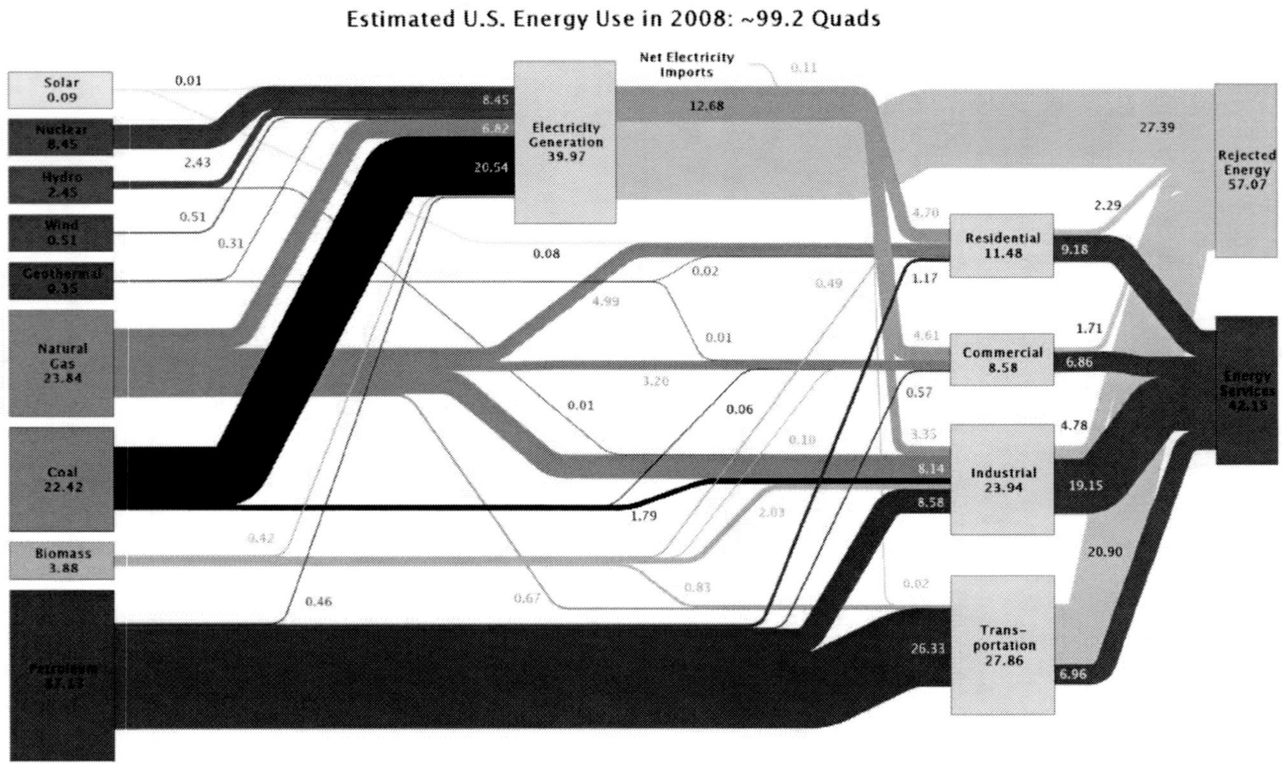

Source: LLNL 2009. Data is based on DOE/EIA-0384(2008), June 2009. If this information or a reproduction of it is used, credit must be given to the Lawrence Livermore National Laboratory and the Department of Energy, under whose auspices the work was performed. Distributed electricity represents only retail electricity sales and does not include self-generation. EIA reports flows for non-thermal resources (i.e., hydro, wind and solar) in BTU-equivalent values by assuming a typical fossil fuel plant "heat rate." The efficiency of electricity production is calculated as the total retail electricity delivered divided by the primary energy input into electricity generation. End use efficiency is estimated as 80% for the residential, commercial and industrial sectors, and as 25% for the transportation sector. Totals may not equal sum of components due to independent rounding. LLNL-MI-410527

Figure 4. U.S. Primary Energy Use in 2008.

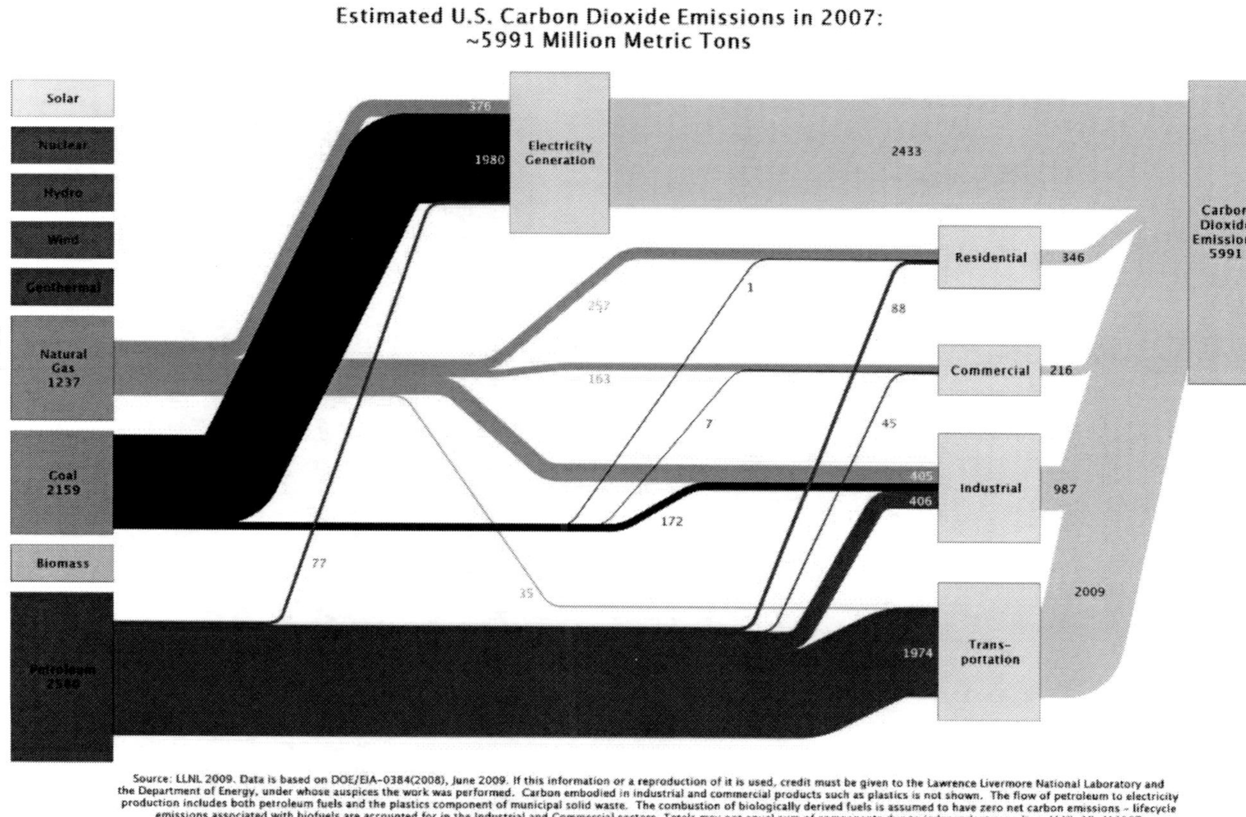

Figure 5. U.S. Carbon Dioxide Emissions in 2007.

The Administration's clean energy and climate change objectives are ambitious and achievable. Successful achievement of these objectives will require solutions to technical challenges associated with various energy sectors, including:

- *Electricity Sector GHG Production* – As seen in Figures 4 and 5, the U.S. electricity production sector annually consumes 40 quadrillion BTU of primary energy, producing 4,150 million MWh of electricity, and emitting 2,400 million MT of CO_2. The average carbon intensity of the U.S. electric-generating sector is 0.58 MT–CO_2/MWh of electricity produced. While far from the world's highest carbon intensity (China produces 0.87 MTCO_2/MWh of electricity), U.S. electric-generating-sector carbon intensity is far higher than some industrialized countries. For instance, France emits only 0.09 MT–CO_2/MWh of electricity produced. There is clearly both the need for, and the real potential for, significant improvement in U.S. electric-generating-sector carbon intensity and GHG emissions.
- *Transportation Sector Energy Use and GHG Emissions* – The transportation sector is currently responsible for 33% of GHG emissions (Figure 5). In addition to more energy-efficient internal combustion engines, electrification of the transportation sector using new low-carbon electricity-generation technologies will assist in reducing these emissions. Successful electrification of the transportation sector is also dependent on improvements in battery technology to enable high-density energy storage to meet vehicle service range requirements.
- *Industrial Sector Energy Use and GHG Emissions* – Industrial use of energy is responsible for 16 percent of the country's GHG emissions (Figure 5). About half of these emissions come from chemical facilities and oil refineries. The development of GHG-free technologies that can generate and deliver significant thermal and chemical energy to industry is needed.

> The driver for the new energy policy is to continue to generate energy, mostly from domestic sources, at an affordable price. The policy must meet increasing demand, with considerably reduced GHG emissions, and without stifling GDP growth.

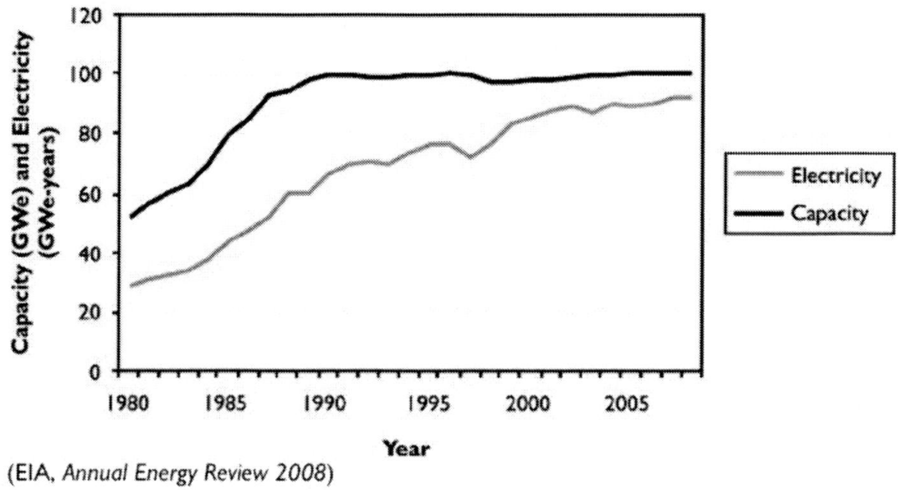

Figure 6. U.S. Nuclear Energy History, 1980 – 2008.

2.2. The Value and Need for an "Energy Portfolio" Approach

Given the issues noted in Section 2.1, an effective energy policy will almost certainly rely on the development and use of a portfolio of domestic clean energy sources. This is true not only because of resource limits at various points in the energy supply chain but also because all energy sources face economic, technical, and societal risks to their successful deployment.[4] R. Socolow and S. Pacala, in "A Plan To Keep Carbon In Check,"[5] have demonstrated the potential for energy portfolio approaches to enhance U.S. energy security and reduce the threat of global warming. The following section discusses the role of nuclear energy as an element of the U.S. energy portfolio.

2.3. Nuclear Energy as an Element of the Future U.S. Energy Portfolio

In 2007, the 104 light-water reactors (LWRs) currently operating in the United States generated 806 billion kilowatt-hours (kW-hrs), equivalent to 92 gigawattyears (GWe-yrs). As shown in Figure 6, even though the generating capacity of the nuclear fleet has been essentially flat for almost twenty years, the production of nuclear electricity continued to grow largely as a result of increased capacity factors. The fleet's average capacity factor improved from 56.3% in 1980 to 91.9% in 2008.[6] This improvement was driven by reactor operators and the efforts of the Electric Power Research Institute (EPRI), spurred by NE-sponsored R&D into high-burnup fuels that allowed utilities to shift from 12-month operating cycles to 18- or 24-month operating cycles that reduced downtime. Additionally, some growth can be attributed to power uprates that increased capacity at existing plants.

While in operation, nuclear power plants do not emit GHGs. Every MWh of electricity produced with nuclear energy avoids the emission of approximately 1.0 MT of CO_2 if the same amount of energy had been generated with conventional coal-fired technologies or approximately 0.6 MT of CO_2 if the energy had been produced with natural gas. Since the per capita electricity consumption in the United States is approximately 14 MWh of electricity per year per person, nuclear energy offers the prospect of avoiding what could otherwise be an annual personal carbon footprint from electricity production of up to 14 MT of CO_2. In addition, nuclear power is dependable. It is available day or night, when the wind is blowing and when it is not. After more than three decades of outstanding safety performance, the public acceptance of nuclear energy has turned in favor of its deployment.[7] However, continued and increased use of nuclear energy faces several key challenges:

- *Capital Cost* – The current fleet of nuclear power plants produces electricity at a very low cost (approximately 2–3 cents/kilowatt-hour) because these plants have already repaid the initial construction investments. However, the capital cost of a large new plant is high and can challenge the ability of electric utilities to deploy new nuclear reactors. Thus, it is important to reduce the capital cost by innovative designs. The introduction of smaller reactors might reduce capital costs by taking advantage of series fabrication in centralized plants and may reduce financial risk by requiring a smaller up-front investment.
- *Waste Management* – At present, no permanent solution to high-level nuclear waste management has been deployed in the United States. Innovative solutions will be required to assure that nuclear waste is properly managed. The Administration has initiated the Blue Ribbon Commission on America's Nuclear Future to conduct a review of policies for managing the back end of the nuclear fuel cycle, including all alternatives for the storage, processing, and disposal of civilian and defense used nuclear fuel and nuclear waste. The results will inform

the Government's process to establish a policy for used fuel and waste management. Ultimately, while the need for permanent waste disposal can never be eliminated, transition to nuclear energy technologies that significantly reduce the production of long-lived radioactive waste – rather than deal with it after it is produced – is a desirable goal.

- *Proliferation Risk* – There is considerable interest in the global expansion of nuclear energy. However, such expansion raises concerns about the proliferation of nuclear weapons, including nuclear explosive devices, stemming from access to enrichment and reprocessing activities that might produce weapons-usable materials. Development of innovative technologies and international policies are essential to prevent nuclear proliferation by nation-states as well as nuclear terrorism by rogue entities. Furthermore, a more robust capability to evaluate and compare proliferation and terrorism risks is needed. In addition, it is in the U.S. interest to engage nations contemplating civil nuclear power for the first time in order to help them develop an indigenous infrastructure designed to deploy the technology in a safe and secure manner.
- *Safety and Reliability* – As existing plants continue to operate and new plants and new types of plants are constructed, it is vital that the excellent safety and reliability record of nuclear energy in the United States be maintained. It is also important that the U.S. share its experience with other countries and work with them to ensure safe operation of their plants.

3. Mission and Goals of the Office of Nuclear Energy

The analysis presented in Section 2 supports the conclusion that increased greenhouse gas-free electricity production is necessary to achieve the transition to a clean-energy economy.

3.1. The Office of Nuclear Energy Mission

The primary mission of NE is to advance nuclear power as a resource capable of meeting the nation's energy, environmental, and national security needs by resolving technical, cost, safety, security, and proliferation resistance, through R&D and demonstrations, as appropriate. Progress in these areas should promote the deployment of fission power systems in a socially acceptable, environmentally sustainable, and economically attractive manner.

Four specific research and development objectives for nuclear energy development outline NE's approach to delivering progress in the areas noted above. The objectives are:

- *R&D Objective 1* – Develop technologies and other solutions that can improve the reliability, sustain the safety, and extend the life of current reactors.
- *R&D Objective 2* – Develop improvements in the affordability of new reactors to enable nuclear energy to help meet the Administration's energy security and climate change goals.
- *R&D Objective 3* – Develop sustainable nuclear fuel cycles.
- *R&D Objective 4* – Understand and minimize the risks of nuclear proliferation and terrorism.

The four objectives are discussed more fully in the following sections.

3.2. Nuclear Energy R&D Objectives and the Role of NE in Achieving Them

This section presents a description of the four R&D objectives and NE's role in making progress in these areas.

3.2.1. R&D Objective 1. Develop Technologies and Other Solutions that Can Improve the Reliability, Sustain the Safety, and Extend the Life of Current Reactors

The existing U.S. nuclear fleet has a remarkable safety and performance record, and today these reactors account for 70 percent of the low GHG-emitting domestic electricity production. Extending the operating lifetimes of current plants beyond sixty years and, where possible, making further improvements in their productivity will generate near-term benefits. Industry has a significant financial incentive to extend the life of existing plants, and as such, activities will be cost shared. Federal R&D investments are appropriate to answer fundamental scientific questions and, where private investment is insufficient, to help make progress on broadly applicable technology issues that can generate public benefits.

The DOE role in this R&D objective is to work with industry and, where appropriate, the Nuclear Regulatory Commission (NRC) to support and conduct the long-term research needed to inform major component refurbishment and replacement strategies, performance enhancements, plant license extensions, and age-related regulatory oversight decisions. The DOE R&D role will focus on aging phenomena and issues that require long-term research and are generic to reactor type.

3.2.2. R&D Objective 2. Develop Improvements in the Affordability of New Reactors to Enable Nuclear Energy to Help Meet the Administration's Energy Security and Climate Change Goals

If nuclear energy is to be a strong component of the nation's future energy portfolio, barriers to the deployment of new nuclear plants must be overcome. Impediments to new plant deployment, even for those designs based on familiar light-water reactor technology, include the substantial capital cost of new plants and the uncertainties in the time required to license and construct them. More advanced plant designs, such as small modular reactors (SMRs) and high-temperature reactors (HTRs), will have additional barriers for deployment. These reactors have characteristics that could make them more attractive than today's technology. SMRs, for example, have the potential to achieve lower proliferation risk and more simplified construction than other designs. The development of next-generation reactors could present lower capital costs and improved efficiencies. These reactors may be based upon new designs that take advantage of the advances in high performance computing while leveraging capabilities afforded by improved structural materials. Industry's role in overcoming the barriers in this area is substantial. DOE supports R&D ranging from fundamental nuclear phenomena to the development of advanced fuels that could improve the economic and safety performance of these advanced reactors. Nuclear power can reduce GHG emissions from electricity production and possibly in co-generation by displacing fossil fuels in the generation of process heat for applications including refining and the production of fertilizers and other chemical products.

3.2.3. R&D Objective 3. Develop Sustainable Nuclear Fuel Cycles

Sustainable fuel cycle options are those that improve uranium resource utilization, maximize energy generation, minimize waste generation, improve safety, and complement institutional measures in limiting proliferation risk. The key challenge for the government in this R&D objective is to develop a suite of options that will enable future decision makers to make informed choices about how best to manage the used fuel from reactors. DOE will conduct R&D in this area to investigate the technical challenges involved with three potential strategies for used fuel management.

- *Once-Through* – Develop fuels for use in reactors that would increase the efficient use of uranium resources and reduce the amount of used fuel for direct disposal for each MWh of electricity produced. Additionally, evaluate the inclusion of non-uranium materials (*e.g.*, thorium) in reactor fuel options that may reduce the long-lived radiotoxic elements in the used fuel that would go into a repository.
- *Modified Open Cycle* – Investigate fuel forms and reactors that would increase utilization of the fuel resource and reduce the quantity of long-lived radiotoxic elements in the used fuel to be disposed (per MWh), with limited separations steps using technologies that substantially lower proliferation risk.
- *Full Recycling* – Develop techniques that will enable the long-lived actinide elements to be repeatedly recycled rather than be disposed. The ultimate goal is to develop a cost- effective and low proliferation risk approach that would dramatically decrease the longterm danger posed by the waste, reducing uncertainties associated with its disposal.

DOE will work to develop the best approaches within each of these tracks to inform waste management strategies and decision making.

3.2.4. R&D Objective 4. Understand and Minimize the Risks of Nuclear Proliferation and Terrorism

It is important to assure that access to the benefits of nuclear power can be enabled while limiting nuclear proliferation and security risks. This goal requires an integrated approach that incorporates simultaneous development of nuclear fuel cycle technology, safeguards and security technologies and systems, new proliferation risk assessment tools, and non-proliferation frameworks and protocols. These activities must be informed by robust assessments that identify potential approaches for limiting risks of specific technologies and nuclear fuel cycle system options. NE will work with other organizations such as the National Nuclear Security Administration (NNSA), the Department of State, the NRC, and others in further defining, implementing and executing this integrated approach. Aspects of this research may help to inform the exploration of concepts such as international fuel service arrangements.

4. An Integrated Nuclear Energy Roadmap

This section presents an objective-focused roadmap to advance nuclear energy technologies. As depicted in Figure 7, the activities described here ultimately "unpack" to a suite of science and technology development activities, many of which will support more than one R&D objective.

The approach incorporates a portfolio of long-term R&D objectives and a balanced focus on evolutionary, innovative, and high-risk–high-payoff R&D in many diverse areas. The organization and coordination of the science and technology thrusts ("Enabling Technologies" in Figure 7) will be a focus of program and strategic planning follow-on implementation plants, but is briefly addressed in Section 5.2 of this document.

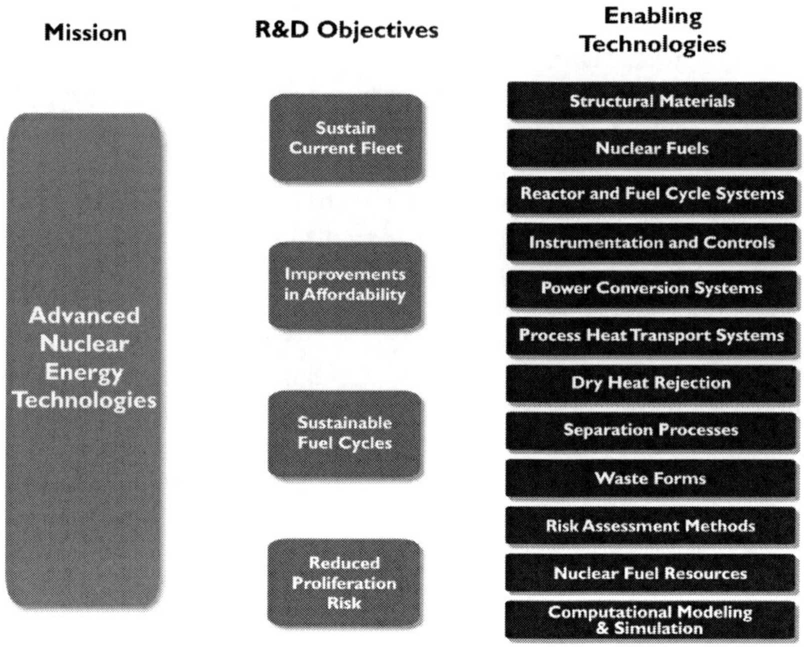

Figure 7. NE Mission, R&D Objectives, and Technologies.

In laying out the activities in each of the R&D objectives described below, we must remain goal- oriented to avoid falling into the trap of doing a great deal of work that, while interesting, fails to address the challenges to the deployment of nuclear energy. The following sections highlight areas in which NE may undertake future R&D. These R&D activities have been considered with the end in mind to ensure that the linkage between research and solution is clear. To that end, in depicting the timelines of activity for the R&D objectives below, the charts show a distinction between near-term milestones toward which the NE R&D plan is designed to progress, represented as triangles, and longer-term potential outcomes that provide a framework for the milestones, which are shown as ovals. The milestone charts attempt to depict the stages of development so as not to leave a sense that new technologies can be immediately deployed at a commercial level. Not every milestone or potential outcome outlined in these charts represent actions that are within DOE's roles and responsibilities, and research paths will include many decision points that require choosing the most promising options for continued R&D. Especially as technology matures, industry has a role and a responsibility to share the costs of making progress. It is ultimately industry's decision which commercial technologies will be deployed. The federal role falls more squarely in the realm of R&D.

These long-term milestones and potential outcomes are not set in stone, and in some cases the following sections outline multiple competing paths within an objective, knowing that ultimately only one direction will be chosen. In all cases, the activities, milestones, and plans outlined in this document will be reconsidered and revised periodically to ensure that NE R&D is consistent with priorities and reflects what we have learned from these efforts. Activities will be reviewed and modified as necessary through the established budgetary and decision-making processes.

Although some smaller component or process "demonstration" activities are mentioned, these are largely field tests and other actions to provide proof or validation of system elements. They are not large-scale demonstrations like the Next Generation Nuclear Plant (NGNP). Any decisions to embark on such large-scale demonstrations will be the result of decision-making processes that include the relevant stakeholders in the Executive Branch and Congress and will be made in accordance with NEPA and DOE Order 413 requirements. This R&D will enable these stakeholders to understand the potential tradeoffs embodied in these decisions.

4.1. R&D Objective 1. Develop Technologies and Other Solutions that Can Improve the Reliability, Sustain the Safety, and Extend the Life of Current Reactors

The current fleet of 104 nuclear power plants has reliably and economically contributed almost 20 percent of electricity generated in the United States over the past two decades. However, by 2030, even those current nuclear power plants that have received 20-year extensions from the NRC authorizing 60 years of life will begin reaching the end of their licensed periods of operation. Figure 8 shows projected nuclear energy contribution to domestic generating capacity from those plants that have already received 20-year license extensions. If current plants do not operate beyond 60 years, the total fraction of generated electricity from nuclear power could begin to decline, even with the addition of new nuclear-generating capacity.

Replacing the current fleet would require hundreds of billions of dollars. Replacement of this 100 GWe-generating capacity with traditional fossil plants would lead to significant increases in CO_2 emissions. Extending operating licenses beyond 60 to perhaps 80 years would enable existing plants to continue providing safe, clean and economic electricity without significant GHG emissions. The objective of this R&D objective is to provide a comprehensive technical basis for extending the life of today's LWRs that could be used to inform licensing and managing the long-term safe and economical operation.

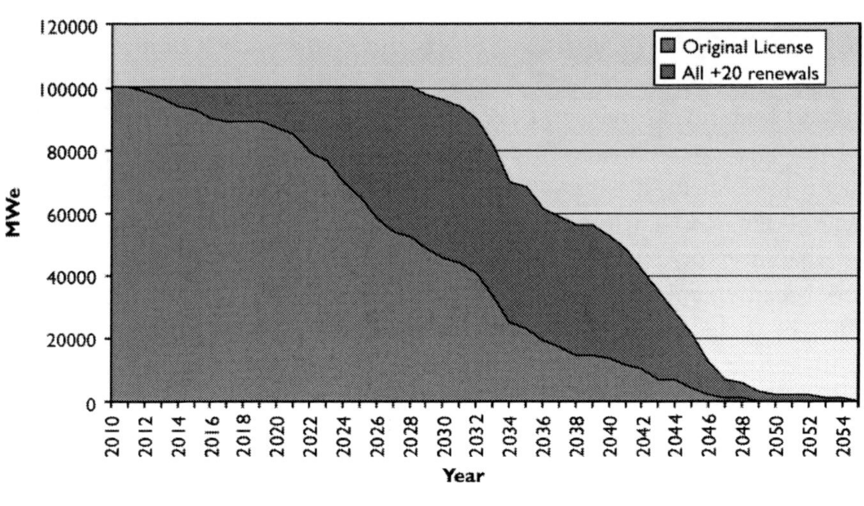

Figure 8. Nuclear Capacity With and Without License Extensions.

4.1.1. Challenges Facing the Current Fleet

The following are the major challenges facing the current fleet:

- Aging and degradation of system structures and components, such as reactor core internals, reactor pressure vessels, concrete, buried pipes, and cables.
- Fuel reliability and performance issues.
- Obsolete analog instrumentation and control technologies.
- Design and safety analysis tools based on 1980s vintage knowledge bases and computational capabilities.

Industry's economic incentive to meet these challenges in order to continue the safe and reliable operation of existing plants is tremendous. As such, federal activities undertaken in this area will be cost-shared with industry. Industry, working through EPRI or through the various owners' groups, will engage some of these problems directly. Federal R&D investments are appropriate to answer fundamental scientific questions and where private investment is insufficient, to help make progress on broadly-applicable technology issues that can generate public benefits. The government holds a great deal of theoretical, computational, and experimental expertise in nuclear R&D that is not available in industry. The benefits of assisting industry with R&D on life-extension apply not only to current plants but also to the next generation of reactor technologies still in development.

4.1.2. R&D Topics for Life Extension and Performance Improvement

The overall focus of the R&D activities will be to improve a power plant operator's ability to manage the effects of the aging of passive components and increase operational efficiency and economics. In selecting projects for federal investment, it is vital that due consideration be given not only to how each of the R&D activities support achievement of safety and economic sustainability for existing LWRs, but also to how the R&D results will be more broadly applicable to the next generation of reactor technologies. These activities should also be integrated with outside sources of information and parallel R&D programs in industry, the NRC, universities, and other laboratories, both domestic and international. Close coordination with the NRC as appropriate is needed to assure that R&D programs focus on issues relevant to licensing.

The following are R&D topics where NE will focus its efforts to help provide solutions to the challenges listed above, thereby helping enable reactor life extension beyond 60 years with improved performance. Progress on this long-term and high-risk–high-reward R&D, which supports the current nuclear power plant fleet, will provide the scientific underpinnings for plant owners to make billion-dollar investment decisions to prolong the economic lifetime of these assets. R&D findings will also inform improvements in the lifetime of future-generation reactor designs.

- *Nuclear Materials Aging and Degradation* – Develop a scientific basis for understanding and predicting long-term environmental degradation behavior of materials in nuclear power plants. Provide data and methods to assess performance of systems, structures, and components essential to safe and sustained nuclear power plant operation.
- *Advanced LWR Nuclear Fuel Development* – Improve the scientific knowledge basis for understanding and predicting fundamental nuclear fuel and cladding performance in nuclear power plants. Apply this information to the development of high-performance, high-burnup fuels with improved safety, cladding, integrity, and economics.
- *Advanced Instrumentation, Information, and Control (II&C) System Technologies* – Research to address long-term aging and obsolescence of existing instrumentation and control technologies and to develop and test new technologies. Establishing a strategy to implement long-term modernization of II&C systems will be the focus of federal R&D, while industry will focus on the more immediate benefits of adapting existing digital technologies to current plants. NE will work with industry to develop advanced condition- monitoring technologies for reliable plant operation, improved understanding of physical methods of degradation, and the means to detect and characterize these processes.
- *Risk-Informed Safety Margin Characterization (RISMC)* – Bring together risk-informed, performance-based methodologies with fundamental scientific understanding of critical phenomenological conditions and deterministic predictions of nuclear plant performance to provide an integrated characterization of public safety margins in aging nuclear power plants. Such an approach will better characterize safety margins and should improve the reliability and efficiency of

plant operations. RISMC will also be applicable to future generations of nuclear power plants.

- *Efficiency Improvement* – Improve the efficiency of the current fleet while maintaining excellent safety performance is one of the primary objectives of life extension. Power uprates have contributed to improving the current fleet's economic performance. This activity focuses on developing methodologies and scientific bases to enable more extended power uprates.
- *Advanced Modeling and Simulation Tools* – Conduct R&D needed to create a new set of modeling and simulation capabilities that will be used to better understand the safety performance of the aging reactor fleet. These tools will be fully three-dimensional, high- resolution, modeling integrated systems based on first-principle physics. To accomplish this, the modeling and simulation capabilities will have to be run on modern, highly parallel processing computer architectures.

The sustainability of light water reactors will benefit enormously from advanced modeling and simulation capabilities. The NE Modeling and Simulation Hub will integrate existing nuclear energy modeling and simulation capabilities with relevant capabilities developed by the Office of Science, the NNSA, and others. The results will leapfrog current technology to provide a multi- physics, multi-scale predictive capability that is a revolutionary improvement over conventional codes. A key challenge will be to adapt advanced computer science tools to an applications environment. The hub is intended to create a new state-of-the-art in an engineering-oriented multi-physics computational environment that can be used by a wide range of practitioners to conduct ultra-high fidelity predictive calculations of reactor performance.

4.1.3. Key Activities

The following chart outlines potential milestones and future national industry aims for this objective. It presents a distinction between near-term milestones toward which the NE R&D plan is designed to progress, represented as triangles, and longer-term potential outcomes that provide a framework for the milestones, shown as ovals. The milestone charts attempt to depict the stages of development so as not to leave a sense that new technologies can be immediately deployed at a commercial level. Not every milestone or potential outcome outlined in these charts represents actions that are within DOE's roles and responsibilities, and research paths will include many decision points that require choosing the most promising options for continued R&D. All DOE R&D activities will be evaluated and revisited regularly and modified as necessary through the budget process to ensure the portfolio reflects past progress and current priorities.

Although some smaller component or process "demonstration" activities are mentioned, these are largely field tests and other actions to provide proof or validation of system elements. They are not costly, large-scale demonstrations like NGNP. Any consideration to embark on such large-scale demonstrations will be the result of decision-making and budget development processes.

4.2. R&D Objective 2. Develop Improvements in the Affordability of New Reactors to Enable Nuclear Energy to Help Meet the Administration's Energy Security and Climate Change Goals

The previous 30-year U.S. hiatus in new nuclear plant orders presents a number of immediate hurdles for the construction of new plant designs. Utility investors are still wary of the new regulatory framework, which will not be fully exercised until the first new plant begins operation. There are also concerns regarding the large capital costs of plants and associated difficulties in financing their construction.

NE's objective is to assist in the revitalization of the U.S. industry through R&D. By advancing technologies through R&D, NE can help accelerate deployment of new plants in the short term, support development of advanced concepts for the medium term, and promote design of revolutionary systems for the long term. Work will be done in partnership with industry to the maximum extent possible. Elements of NE's strategy in this area include:

- Assist industry to improve light water reactors using existing technologies and designs.
- Explore advanced LWR designs with improved performance.
- Research and develop small modular reactors that have the potential to achieve lower proliferation risks and more simplified construction than other designs.
- In the longer term, support R&D of advanced reactor technologies that offer lower costs and waste generation.

- Investigate revolutionary reactor concepts that promise to significantly reduce costs and improve performance of nuclear energy.
- Support R&D of nuclear energy's potential to displace fossil fuels in the production of process heat.

Implementing this strategy will require that DOE work in partnership with the nuclear industry and, to the degree appropriate, the NRC.

4.2.1. Challenges Facing New Reactor Deployments

There are several new plant designs, often referred to as Gen III+, that have been certified or are being reviewed by the NRC for immediate deployment in the United States. Potential owners of these Gen III+ plants must overcome serious financial hurdles. All near-term options for new plants are large LWR designs that are optimized for baseload electricity production. Smaller reactors that could be deployed in modules might help reduce the up-front capital costs associated with large plants by allowing utilities to incrementally "step up" to larger electrical capacities while generating revenue and repaying initial debts. New reactor designs beyond Gen III+ may also be deployed. In many cases, new technologies will be needed to enable these new designs, and innovative features will need to be fully demonstrated. Certain aspects of the regulatory framework need to accommodate these new technologies and design features, especially for designs that differ significantly from the large LWR plants in operation today. Economic competitiveness will remain the major hurdle for all novel concepts, including smaller reactors and reactors for non-electric applications.

During the 30-year hiatus from new plant orders in the United States, some nations have continued to grow their nuclear industries. As a result, some other countries have advanced the state-of-the art in manufacturing of nuclear plant components and have made progress in applying more efficient construction techniques. The domestic industry can learn from these international experiences.

4.2.2. R&D Topics for Enabling New Builds

In the United States, it is the responsibility of industry to design, construct, and operate commercial nuclear power plants. However, DOE has statutory authority under the Atomic Energy Act to promote and support nuclear energy technologies for commercial applications. In general, appropriate government roles include researching high-potential technologies beyond the investment horizon of industry and also reducing the technical risks of new technologies. In the case of new commercial reactor designs, potential areas of NE involvement could include:

- Enabling new technologies to be inserted into emerging and future designs by providing access to unique laboratory resources for new technology development and, where appropriate, demonstration.
- Working through the laboratories and universities to provide unique expertise and facilities to industry for R&D in the areas of:
    - Innovative concepts and advanced technologies.
    - Fundamental phenomena and performance data.
    - Advanced modeling and simulation capabilities.
    - New technology testing and, if appropriate, demonstration.
    - Advanced manufacturing methods.

Representative R&D activities that support each of the roles stated above are presented below. The level of DOE investment relative to industry investment will vary across the spectrum of these activities, with a generally increasing trend in DOE investment for longer-term activities. Finally, there is potential to leverage and amplify effective U.S. R&D through collaborations with other nations through multilateral and bilateral agreements including the Generation IV International Forum, which is investigating multiple advanced reactor concepts. DOE is also a participant in OECD/NEA and IAEA initiatives that bear directly on the development and deployment of new reactor systems.

4.2.2.1. Accelerate Advancements in LWR Designs

Given the maturity of the Gen III+ LWR designs, R&D needs are necessarily limited, as the design of these plants is well underway or already complete, some of them are being built overseas, and many have been ordered in the United States and elsewhere. Nevertheless the R&D topics identified jointly with industry for R&D Objective 1 are all applicable to this task.

R&D of more advanced LWR concepts, including novel materials, fuels, and innovative system architectures, is a legitimate role for DOE and its laboratories in partnership with industry. This R&D will help address long-term trends in the capital cost of large LWR plants. Much of this research is also expected to be applicable to non-LWR technologies.

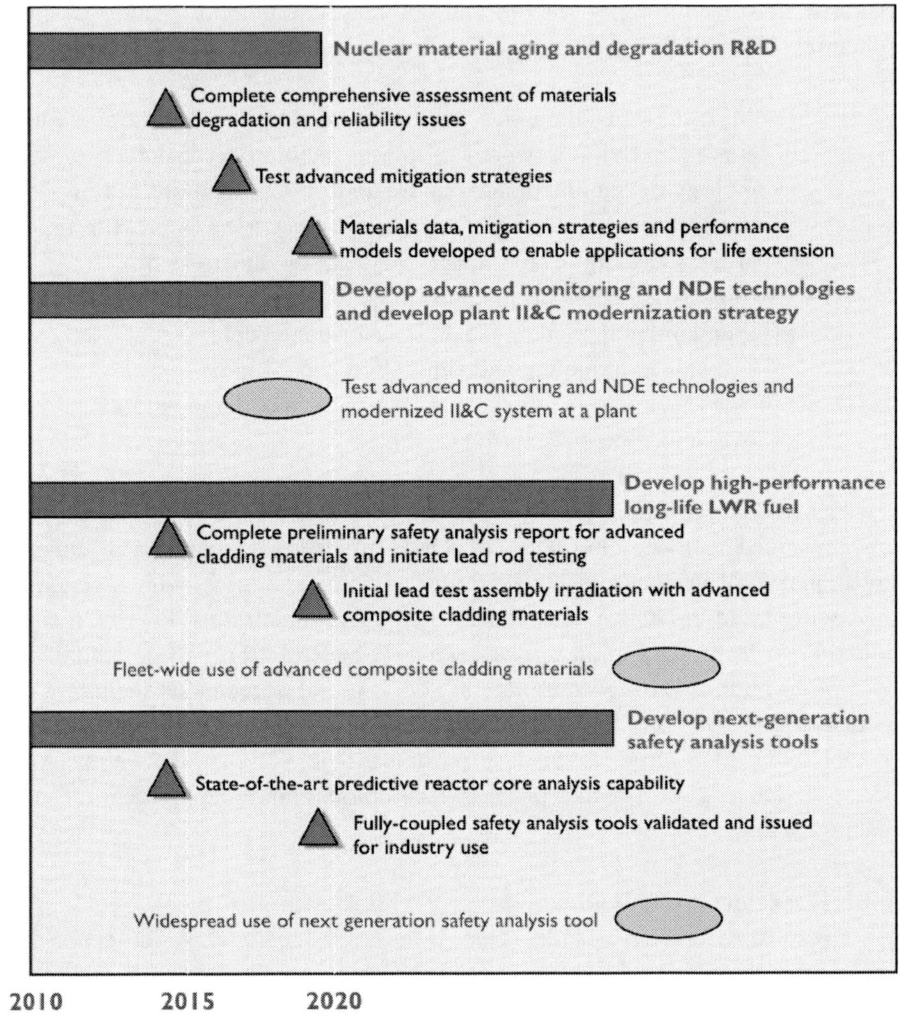

Figure 9. Key Activities for R&D Objective 1.

4.2.2.2. Accelerate the Development of SMR Designs

Several U.S.-based companies are seeking to bring new SMR designs to market, including some with potential for deployment within the next decade. Many of these designs use well- established light-water coolant technology to the fullest extent possible to shorten the timeline for deployment. As such, R&D needs for these technologies are minimal. However, these designs may include new features, such as the use of an integral primary system reactor (IPSR) design and components that are not currently used in commercial plants, such as helical-coil steam generators. DOE will hold workshops with LWR SMR vendors and suppliers, potential utility customers, national laboratory and university researchers, DOE, NRC, and other stakeholders to identify potential priorities to enable their commercialization and development. The Administration will evaluate potential priorities in the context of the appropriate federal role to identify the most cost-effective, efficient, and appropriate mechanisms to support further development.

SMR designs that are not based on LWR technology have the potential to offer added functionality and affordability. In this area, NE will support a range of R&D activities, such as basic physics and materials research and testing, state-of-the-art computer modeling and simulation of reactor systems and components, probabilistic risk analyses of innovative safety designs and features, and other development activities that are necessary to establish the concept's feasibility for future deployment. For SMRs that are based on concepts with lower levels of technical maturity, the Department will first seek to establish the R&D activities necessary to prove and advance innovative reactor technologies and concepts. The Department will support R&D activities to develop and prove the proposed design concepts. Emphasis will be on advanced reactor technologies that offer simplified operation and maintenance for distributed power and load-following applications and increased proliferation resistance and security.

Activities will focus on showing that SMRs provide an innovative reactor technology that is capable of achieving electricity generation and performance objectives that meet market demands and are comparable, in both safety and economics, to the current large baseload nuclear power plants.

NE may also support the development of new/revised nuclear industry codes and standards necessary to support licensing and commercialization of innovative designs and, consistent with NRC guidance and regulations, identify activities for DOE funding to enable SMR licensing for deployment in the United States.

4.2.2.3. Develop Advanced Reactor Technologies

Future-generation reactor systems will employ advanced technologies and designs to improve performance beyond what is currently attainable. Moving beyond LWR technology, for example, may enable reactors to operate at higher temperatures and improved efficiencies resulting in improved economics. Advanced materials may make reactors easier to construct while also enabling better performance. Improved designs utilizing these advances could reduce the capital costs associated with the current set of reactors being considered. Two prominent examples of advanced reactor technologies worthy of further investigation include:

- The high temperature gas-cooled reactor (HTGR), a graphite moderated thermal-spectrum reactor operated at high temperature for efficient generation of electricity and heat delivery for non-electric applications.
- Fast-spectrum reactors that could provide options for future fuel cycle management and could also be used for electricity generation (see R&D Objective 3).

The U.S. is also a member of the Generation IV International Forum, which is investigating additional advanced reactor systems that employ comparatively less mature technologies while offering significant potential for performance, safety, and economic advances.

Key areas of R&D for future systems could include:

- High-performance materials compatible with the proposed coolant types and capable of extended service at elevated temperatures.
- New fuels and cladding capable of irradiation to high burnup.
- Advanced heat delivery and energy conversion systems for increased efficiency of electricity production.
- Advanced modeling and simulation tools that can reduce uncertainties in predicted performance, improve characterization of uncertainties, and streamline the design of new reactor technologies.
- Systems design for revolutionary new reactor concepts.

4.2.2.4. Develop Technologies Consistent with Both Electric and Non-Electric Applications

An additional potential benefit from nuclear power could be realized through new plant designs that would be used to displace GHG-emitting fuels in the industrial sector while also generating electricity. Some industrial process heat applications require temperatures substantially above the 300–325°C outlet temperature of today's LWRs. Petroleum refining, for example, requires temperatures in the range of 250-500°C while steam reforming of natural gas requires process heat in the 500-900°C range. Achieving higher output temperatures requires switching to a new coolant technology such as gas, liquid metal, or molten salt. With these coolants, it may be possible to achieve outlet temperatures ranging from over 500°C for liquid metal coolants to over 900°C for helium or molten salt coolants. Achieving these temperatures, however, will require the development and qualification of fuels, materials and instrumentation, particularly at the higher end of the temperature range. Also, the use of coolants other than water will require the development of a variety of plant components and systems such as electromagnetic pumps for liquid metal coolants, compact heat exchangers for gas coolants, and chemical purification systems for molten salt coolants. These coolants will also require the development of new licensing requirements and codes and standards. While the economic market for dedicated process heat from nuclear power may be limited, reactors that could produce electricity as well as industrial process heat may have broader applications.

Key areas of R&D for future systems could include:

- *Develop interfacing heat transport systems* – Supply process heat with minimal losses to industrial users within several kilometers of the reactor.
- *Develop modeling and simulation capabilities* – These tools would improve understanding of interactions between the kinetics of the various reactor types and the kinetics of the chemical plants or refineries, which they would serve. Modeling may also be used to understand the long-term performance of catalysts and solid-oxide cells at an atomistic level.

4.2.3. Key Activities

The following chart outlines potential milestones and future national industry aims for this objective. It presents a distinction between near-term milestones toward which the NE R&D plan is designed to progress, represented as triangles, and longer-term potential outcomes that provide a framework for the milestones, shown as ovals. The milestone charts attempt to depict the stages of development so as not to leave a sense that new technologies can be immediately deployed at a commercial level. Not every milestone or potential outcome outlined in these charts represents actions that are within DOE's roles and responsibilities, and research paths will include many decision points that require choosing the most promising options for continued R&D. All DOE R&D activities will be evaluated and revisited regularly and modified as necessary through the budget process to ensure the portfolio reflects past progress and current priorities.

Although some smaller component or process "demonstration" activities are mentioned, these are largely field tests and other actions to provide proof or validation of system elements. They are not costly, large-scale demonstrations like NGNP. Any consideration to embark on such large-scale demonstrations will be the result of decision-making and budget development processes.

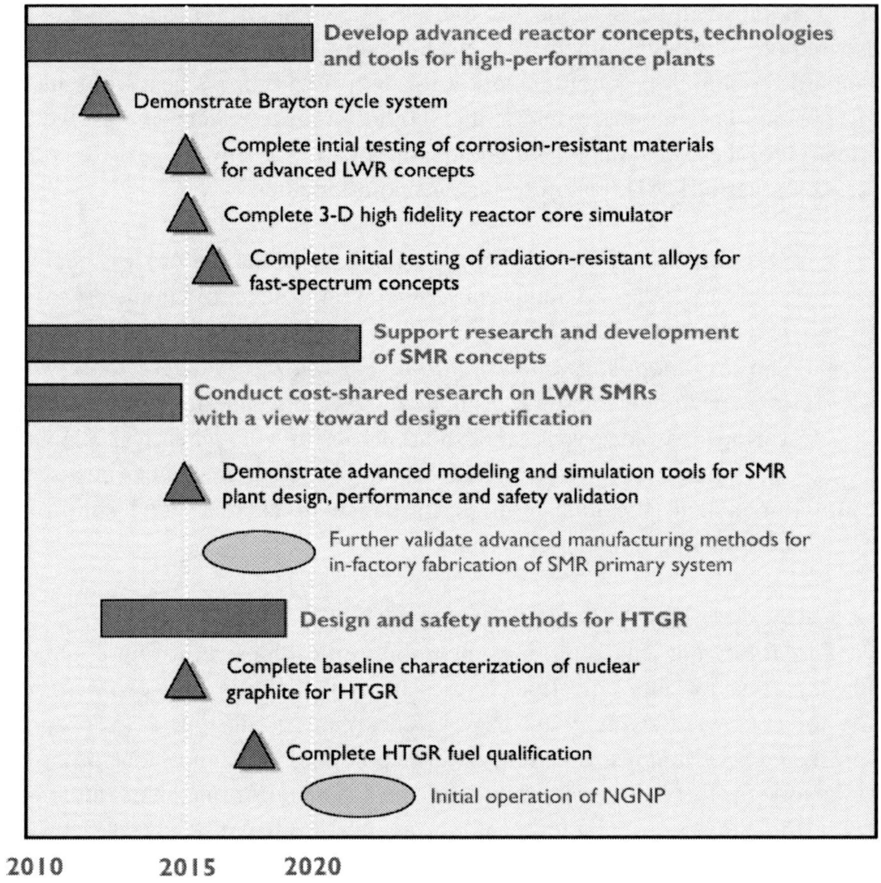

Figure 10. Key Activities for R&D Objective 2.

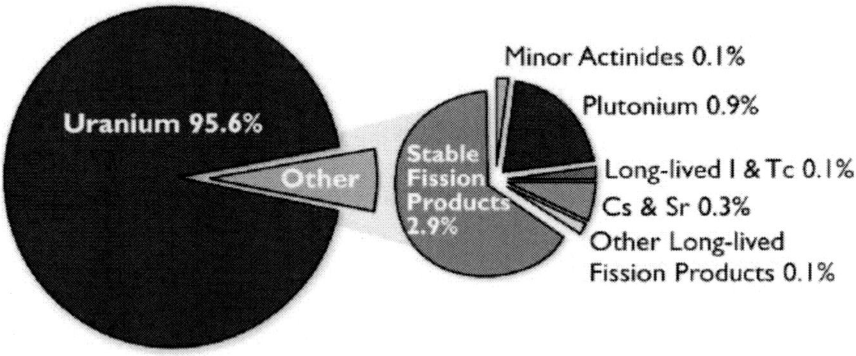

Figure 11. Constituents of Used LWR Fuel.

4.3. R&D Objective 3. Develop Sustainable Nuclear Fuel Cycles

Sustainable fuel cycle options are those that improve uranium resource utilization, maximize energy generation, minimize waste generation, improve safety, and limit proliferation risk. The principal challenge for the government in this objective is to develop a suite of options that will enable future decision makers to make informed choices about how best to manage the used fuel from reactors. The Administration has established the Blue Ribbon Commission on America's Nuclear Future to inform this waste management decision-making process. The Commission will review policies for managing the back end of the fuel cycle including alternatives for the storage, processing, and disposal of civilian and defense used nuclear fuel and nuclear waste. All research and development activities and plans outlined here will be revisited and revised as needed to reflect the Commission's findings and associated Administration decisions.

An expansion of nuclear power in the United States will result in a growth of the used nuclear fuel inventories. The Nuclear Waste Policy Act of 1982 gave the U.S. government the mission to safely manage the used fuel from these nuclear power plants. Research and development of sustainable nuclear fuel cycles and waste management activities is important to support the expansion of nuclear energy. Some of the attributes of the sustainable fuel cycle, including waste management and disposal technologies, include the responsible use of natural resources, preservation of the environment for future generations, safety, security, public acceptance, and cost effectiveness.

The constituents of current used nuclear fuel (UNF) after discharge from LWRs are shown in Figure 11. As this figure shows, the vast majority of the material in the used fuel is uranium that is generally unchanged from the fuel that went into the reactor to produce energy. Uranium is considered an element in the category called "actinides," along with the "transuranic" elements of plutonium and the "minor" actinides: neptunium, americium, and curium, principally. These elements generally are long-lived and must be isolated from the environment for tens or hundreds of thousands of years. Actinides are also of interest because uranium and plutonium could be recycled to produce more energy in reactors, as could the minor actinides in fast-spectrum reactors. The remaining class of elements in the used fuel is fission products, many of which are stable and pose little concern. The short- lived fission products – primarily cesium and strontium – generate most of the hazard for the first hundreds of years of disposal. There are also fission products, notably iodine and technetium, that last for hundreds of thousands of years and must be isolated from the environment.

NE will research and develop nuclear fuel and waste management technologies that will enable a safe, secure, and economic fuel cycle. The NE R&D strategy will be to investigate the technical challenges that would be encountered in each of three potential methods and perform R&D within each of these tracks:

- *Once-Through* – Nuclear fuel makes a single pass through a reactor after which the used fuel is removed, stored for some period of time, and then directly disposed in a geologic repository for long-term isolation from the environment. The used fuel will not undergo any sort of treatment to alter the waste form prior to disposal in this approach, eliminating the need for separations technologies that may pose proliferation concerns. Less than one percent of the mined uranium is utilized in the present once-through fuel cycle.
- *Modified Open Cycle* – The goal of this approach is to develop fuel for use in reactors that can increase utilization of the fuel resource and reduce the quantity of actinides that would be disposed in used fuel. This strategy is "modified" in that some limited separations and fuel processing technologies are applied to the used LWR fuel to create fuels that enable the extraction of much more energy from the same mass of material and accomplish waste management goals.
- *Full Recycle* – In a full recycle strategy, all of the actinides important for waste management are recycled in thermal- or fast-spectrum

> systems to reduce the radiotoxicity of the waste placed in a geologic repository while more fully utilizing uranium resources. In a full recycle system, only those elements that are considered to be waste (primarily the fission products) are intended for disposal, not used fuel. Implementing this system will require extensive use of separation technologies and the likely deployment of new reactors or other systems capable of transmuting actinides.

The R&D approach will be to understand what can be accomplished in each of these strategies and then to develop the promising technologies to maximize their potential. One element that crosscuts all potential approaches is disposal and R&D activities will include a focus on those technologies. Additionally, storage will be an important part of any strategy, and R&D will be needed to assess the performance of storage technologies with higher-burnup used LWR fuels, as well as any potential new fuels that may be deployed in the future.

The discussion above is primarily focused on the uranium fuel cycle that is the norm throughout the world. An alternative that could be considered would be the use of thorium to replace at least part of the uranium in the system. Thorium could be used as part of a once-through, modified open, or full recycle fuel cycle. The appeal of thorium is two-fold. First, thorium is more abundant in nature than uranium and can be used to extend or replace uranium in the fuel cycle. Second, the use of thorium enables reduced production of transuranic elements that end up in used fuel. However, there are still technical and economic challenges facing thorium-based fuels. Thus some R&D to address related challenges may be considered. Significant R&D in the use of thorium has been performed previously in the United States and is currently being considered in other parts of the world (particularly in India).

Unlike R&D Objectives 1 and 2, management of UNF and development of fuel cycle technologies are primarily the government's responsibilities because the government is legally responsible for UNF. Thus, the necessary research, development, and demonstration, if appropriate, will be led primarily by the government. However, early and continuous industrial involvement is important because any technologies that are developed will ultimately be implemented by the commercial entities.

4.3.1. Major Challenges Associated with Fuel Cycle Options

Each of the potential fuel cycle strategies faces challenges, some of which may be shared with other approaches. Similarly, the R&D needed to overcome these challenges may support more than one strategy.

- *Once-Through* – Improving the sustainability of a once-through approach to used fuel management begins with increasing the burnup of the fuel – the amount of energy that can be extracted from fuel in the reactor – which may also have the effect of consuming more actinides in the fuel, leaving less to be disposed. Increasing the burnup of a fuel will require ensuring that both the fuel itself and the structural material designed to keep it in place in the reactor will be able to withstand extended irradiation in the reactor and maintain its integrity when being stored after removal. Deploying advanced fuels will require that they first undergo a qualification process that can take a great deal of time, as researchers must irradiate and conduct examinations on test samples to assure their performance. Also, fuels that are notably different from those currently used in LWRs may drive changes in the fuel processing infrastructure that has evolved to meet current needs. To the extent that the deployed once-through fuel cycle is built upon enriched uranium fuels, the proliferation concerns associated with enrichment technologies will need to be addressed.
- *Modified Open Cycle* – A modified open cycle faces some of the same challenges as the once-through, along with some encountered in a full recycle approach. The modified open cycle introduces the possibility of a used fuel separations step to enable more options for producing fuels. This flexibility enables the inclusion of transuranic elements – notably plutonium – at concentrations capable of supporting ultra-high burnup, along with the attendant difficulties of developing these fuels. The challenges of developing high-burnup fuels discussed in the previous paragraph are applicable to this strategy. The use of separations technology to prepare the ultra-high-burnup fuel introduces difficulties in separations as well as managing proliferation concerns. A key element of this fuel cycle is the likely need to introduce advanced reactors that can utilize these new fuels. The overarching challenge in making a modified open cycle worthwhile is to determine if the improvement in fuel resource utilization and in the waste to be disposed is sufficient to justify the additional

complication, potential proliferation concerns, and expense this approach would entail.

- *Full Recycle* – In a full recycle approach, used fuel is not directly disposed in a repository; rather, those elements of the used fuel that are deemed appropriate for recycling are reintroduced into reactors or other systems while the remaining elements are stabilized in a waste form and disposed. This strategy offers the potential of waste forms that pose far less long-term concern, although the approach would require overcoming not only technical challenges but also economic, proliferation, and public perception concerns. This system would rely on multiple separations processes that must minimize process losses and waste generation while addressing proliferation concerns. Furthermore, fuels must be developed that will allow for the inclusion of all of those elements that are to be recycled in concentrations that vary over time. This is a central tradeoff in the full recycle approach: the more elements that are recycled, the better the waste form will be; however, more separation of elements in the fuel increases the technical and other challenges. Elements that are recycled must be capable of transmutation in a system – likely, but not necessarily, a fast reactor – to eventually eliminate them. In order for a full recycle strategy to be considered, the waste benefits and improved resource utilization produced by such a system must outweigh the complication, expense and potential proliferation concerns associated with it.

4.3.2. R&D for Sustainable Fuel Cycle Options

There are major R&D needs to understand how best to overcome the challenges posed by each of the fuel cycle approaches being considered. The potential R&D efforts that DOE would undertake would have a long-term view and would be science-based. It would take considerable time before the issues in the modified open and the full recycle alternatives would be overcome. Many R&D areas will be applicable to multiple strategies. Prior to beginning major R&D work in these areas, analyses will be performed to gauge the likely value of the efforts.

- *Fuel Resource Exploration and Mining* – The availability of fuel resources for each potential fuel cycle and reactor deployment scenario must be understood. Extended use of nuclear power may drive improvements in defining resource availability and on fuel resource exploration and mining. Primarily, this is work that the private sector would undertake, and how and when this would occur would depend on price and other market conditions. This is most relevant for a once-through approach, but even modified open cycles and full recycle systems may require comparable levels of natural sources of fuel for the foreseeable future. Most appropriate for federal involvement in this area would be R&D to support investigation of long-term, "game-changing" approaches such as recovering uranium from seawater.

- *Used Fuel Disposition* – All radioactive wastes generated by existing and future fuel cycles will need to be safely stored, transported, and disposed. This R&D will identify options for performing these functions, including research into disposal in a variety of geologic environments. This R&D will consider used fuel and high-level waste inventories arising from the current reactor fleet and any additional new builds, including the potential for changing used fuel characteristics from enhanced operations (*e.g.*, increased fuel burnup) and the projected inventories from advanced reactor and fuel cycle systems (*e.g.*, HTRs and SMRs). This research is important to all of the potential fuel cycle approaches.
- *Reduce Transuranic Production In Reactors* – One thrust in developing sustainable fuel cycles will be the exploration of nuclear fuels and reactors that significantly reduce the long-lived actinide content of the used fuel per MWh of energy produced. Exploration of avenues both to reduce actinide production in present and near-term LWRs and to develop future non-LWR systems that produce lower actinide inventories in their used fuel is important. This research area is central to developing the high burnup fuels that will improve the attractiveness of a once-through or modified open fuel cycle.
- *Separation and Partitioning* – The development of processes to recycle used fuel is needed, as well as an evaluation of the feasibility and risks associated with recycling. The objective is to use a predictive approach to evaluate separation chemistry and processes to achieve the desired performance in terms of product purity, environmental impact, and losses. Though not applicable in a once-through system, this topic would be germane to a modified open cycle approach and central to a full recycle strategy.
- *Waste Forms* – It is necessary to develop understanding of waste form behavior over time to help inform decisions on recycle and disposal

options. This understanding must extend over a broad range of potential waste chemistry and disposal environments so waste forms can be adapted and implemented when specific repository conditions are known. This R&D area may be somewhat relevant to strategies that rely on the direct disposal of certain used fuels (such as disposal of high-temperature gas reactor fuels) but the development of improved waste forms is a key component in enabling a full recycle strategy to achieve its promise.

- *Fuel Forms* – The science-based approach will combine theory, experiments, and multi- scale modeling and simulation aimed at a fundamental understanding of the fuel fabrication processes and fuel and clad performance under irradiation. The objective is to use a predictive approach to design future fuels and cladding to enable the development of ultrahigh-burnup fuels in a modified open cycle and to demonstrate the inclusion of recovered actinides in transmutation fuels under a full recycle approach. In the early phases of the program, the major fuel fabrication activities include development of innovative processes to enhance the process efficiency and to improve the control of fuel microstructure for enhanced performance, including tailored fuel forms designed to limit excess actinides across the complex.
- *Material Reuse* – The research will focus primarily on recovered uranium for reuse in reactors to obviate the need to dispose of this material once separated from the rest of the used fuel. The critical areas that require process or equipment modifications will be identified, and technologies will be developed to enable the reuse (and in some cases the re-enrichment) of recycled uranium. Efforts will also investigate the potential recycling and reuse of other constituents of used fuel, such as the zirconium cladding, that are potentially useful but not currently being considered by industry because of uncertainties about material characteristics.
- *Transmutation Systems* – Transmutation is a process to change the characteristics of waste by turning recycled elements into elements with more desirable disposal characteristics. While the focus of most recent work has been on fast-spectrum transmutation reactors, thermal-spectrum transmutation can offer some waste management benefits. R&D would focus on broadly applicable issues including areas such as materials and energy conversion. In addition, studies may be conducted to review the technical and economic aspects of

external neutron source-driven transmutation systems to inform whether future investigation in this approach is warranted.

4.3.3. Key Activities

NE's science-based R&D program will provide a more complete understanding of the underlying science supporting the development of advanced fuel cycle and waste management technologies and, therefore, help provide a sound basis for future decision making. The program will also conduct scientific research and technology development to enable storage, transportation, and disposal of used nuclear fuel and all radioactive wastes generated by existing and future nuclear fuel cycles. Over the next decade, the R&D program will mainly be geared to ensuring that the needed breakthroughs and advancements are available and ready when needed. Examples of such technologies would include ultra-deep-burn LWR, HTR, or fast reactor fuel; reactor technologies to support optimized once-through fuel cycles; and advanced fast reactor concepts to support closed fuel cycles. These technologies would encompass all of the known and anticipated advances that could be expected to be available in areas including materials, design methods, components, and energy conversion.

In keeping with Secretary Chu's vision of using science to provide technological breakthroughs to solve America's grand challenges, the program will include long-term, high-risk–high-payoff R&D. This part of the program will seek revolutionary and transformational breakthroughs in systems, materials and components of the fuel cycle that can better meet the program's objectives. Examples of this could include novel reactor concepts such as molten-salt fuel reactors or thorium fuel cycles. Thus while evolutionary advancements are being made, revolutionary advancements will also be pursued such that, if successful, they could replace all or part of existing or near-term technologies. The roadmap includes milestones for selection of technologies as the program matures. Each approach has a set of reference technologies associated with these milestones:

- *Once-Through* – Develop higher-burnup fuel for LWRs.
- *Modified Open Cycle* – Develop ultra-high-burnup fuel for high-temperature gas-cooled reactors using transuranic elements from used LWR fuel. It is assumed that the NGNP or a comparable reactor will be available for fuel testing. Alternative approaches may require access to a fast-spectrum test reactor and nuclear fuel research capabilities.

- *Full Recycle* – Develop technologies to allow repeated recycling of transuranic elements in fast-spectrum reactors. The initial fuel for the fast reactors will come from separated used LWR fuel with successive reloads made from used fast reactor fuel. Access to a fast- spectrum test reactor will be essential for this research, as will nuclear fuel research capabilities.

The following chart outlines potential milestones and future national industry aims for this objective. It presents a distinction between near-term milestones toward which the NE R&D plan is designed to progress, represented as triangles, and longer-term potential outcomes that provide a framework for the milestones, shown as ovals. The milestone charts attempt to depict the stages of development so as not to leave a sense that new technologies can be immediately deployed at a commercial level. Not every milestone or potential outcome outlined in these charts represents actions that are within DOE's roles and responsibilities, and research paths will include many decision points that require choosing the most promising options for continued R&D. All DOE R&D activities will be evaluated and revisited regularly and modified as necessary through the budget process to ensure the portfolio reflects past progress and current priorities.

Although some smaller component or process "demonstration" activities are mentioned, these are largely field tests and other actions to provide proof or validation of system elements. They are not costly, large-scale demonstrations like NGNP. Any consideration to embark on such large-scale demonstrations will be the result of decision-making and budget development processes.

4.4. R&D Objective 4. Understanding and Minimizing the Risks of Nuclear Proliferation and Terrorism

The final R&D objective for nuclear energy is to enable secure nuclear energy expansion by developing and demonstrating options that limit proliferation and physical security risks associated with nuclear power while also achieving economic, public health and safety, and environmental goals. These risks include not only the possibility that nations may attempt to use nuclear technologies in pursuit of a nuclear weapon, but also the concern that terrorists might seek to steal material that could be used in a nuclear explosive device. This requires NE advocacy for, and participation in, an integrated program to develop technologies, frameworks, and policy options for the future nuclear enterprise, cutting across all aspects of the fuel cycle.

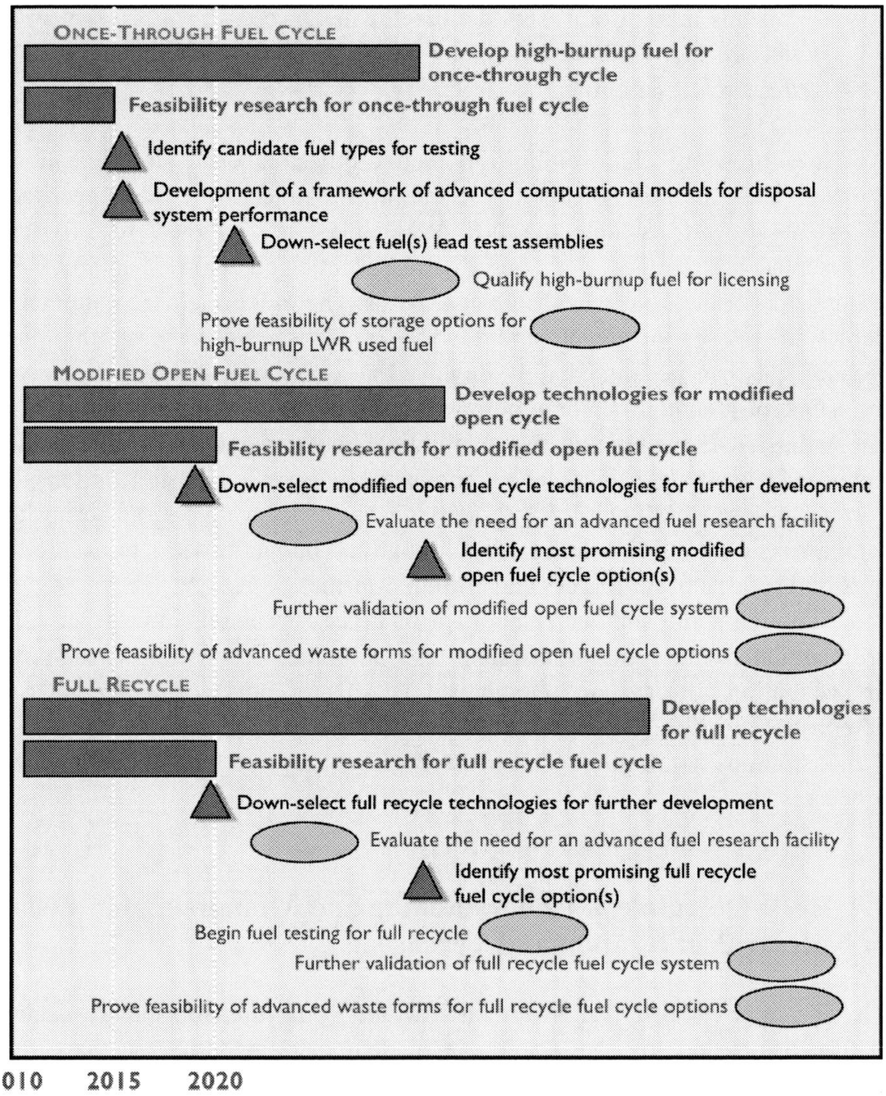

Figure 12. Key Activities for R&D Objective 3.

The United States has extensive experience protecting nuclear materials, from the weapons program that has produced significant quantities of plutonium-239 and highly enriched uranium, to 104 commercial reactors in the U.S. today that handle, use, and store nuclear materials. Internationally, the U.S. has also contributed extensively to the development of technologies now used in the application of international safeguards to monitor used fuel recycling activities in England, France, and Japan. Going forward, safeguards and physical security will become even more integral components in the domestic and global expansion of nuclear power, including the development of future fuel cycle and reactor technologies that further increase the barriers against proliferation and nuclear terrorism.

An integrated U.S. safeguards program provides an opportunity to design improved safeguards and physical security directly into the planning and deployment of new energy systems and fuel cycle facilities. Incorporating safeguards and physical security into the early design phase for new facilities will allow the international community to monitor and verify nuclear material more effectively and efficiently.

DOE has three programs that are collaborating to address safeguards and nonproliferation challenges. The NE Fuel Cycle R&D Materials Protection, Accounting, and Control for Transmutation (MPACT) campaign develops advanced nuclear material management technologies and methods in support of the future domestic U.S. nuclear fuel cycle. The Next Generation Safeguards Initiative (NGSI) within the NNSA Office of Nonproliferation and International Security is designed to leverage U.S personnel, technology, and R&D to add new capacity and significantly strengthen international nuclear safeguards. The third program, the NNSA Office of Nonproliferation Research and Development's Global Nuclear Safeguards R&D Program, whose mission is to support long-term nonproliferation R&D, rounds out the U.S. safeguards R&D efforts for nuclear energy. The work described in this section reflects NE's aspect of the integrated safeguards and nonproliferation work being performed within DOE. This work will be performed in direct collaboration or close coordination with NNSA activities.

In addition to addressing technical safeguards R&D needs, successful integration of these programs would develop revolutionary new tools for proliferation risk assessments and subsequent optimization of advanced nuclear energy systems from nonproliferation and physical security perspectives. The ultimate goal of this crosscutting effort would be to develop and use new analytical tools that could revolutionize our ability to compare proliferation and physical security risk of nuclear energy system options, including aspects of policy and human behavior as well as technical attributes.

As civilian nuclear power expands across the globe, it becomes more important that high standards of safety and security be implemented around the world. Looking only at how the R&D can improve nuclear technologies without considering who is to use these technologies, and the national and international frameworks under which they are operating, will provide an overly narrow perspective of proliferation risks. NE, in cooperation with other DOE offices and national agencies and in partnership and collaboration with other nations, must implement collaborative programs with civilian nuclear power programs in both experienced and inexperienced states in order to minimize proliferation and physical security risks, enhance reactor safety, maximize resource utilization through cooperative R&D, and encourage methods to minimize the dispersion of enrichment and reprocessing facilities worldwide.

4.4.1. Challenges

A key challenge facing the expanded use of nuclear energy and associated fuel cycles is minimizing the potential for the misuse of the technology and materials for weapons purposes. International treaties such as the Nuclear Nonproliferation Treaty, combined with transparency in the use of technology and materials, provide the basic building blocks to assure the peaceful use of nuclear energy. Fuel cycle infrastructure built upon these tenets while enabling the economic provision of fuel cycle services can help prevent the spread of sensitive nuclear technology and materials.

Today's key challenges are to take the wealth of knowledge and experience that exists within the international safeguards and physical security communities and to deploy advanced, affordable techniques to immediately detect the diversion of nuclear materials or the modification of systems. The key technical challenges that must be addressed include:

- Incorporation of nuclear safeguards and physical security technology into designs for fuel cycle facilities, advanced fast reactors, and associated nuclear materials storage and transportation systems.
- Development of proliferation risk assessment methodologies and tools that allow for an integrated view of fuel cycle options to be studied, optimized, and compared.
- Development of advanced containment and surveillance, smart safeguards information management systems, nuclear facility use-

control systems, and next-generation nondestructive analysis and process-monitoring systems.

- R&D of advanced material tracking methodologies, process-control technologies, and plant engineering.
- Remote sensing, environmental sampling, and forensic verification methods.

Addressing these challenges will enable the use and expansion of nuclear energy for peaceful purposes to proceed in a safe and secure manner.

4.4.2. R&D for Understanding and Minimizing the Risks of Nuclear Proliferation and Terrorism

Some potential R&D areas for Objective 4 are:

- *Proliferation Risk Assessments* – Any fuel cycle technologies deployed in the U.S. must be considered in light of how other nations might choose to incorporate them into their own nuclear enterprises. Towards this end, it is important for NE to develop a means of understanding how these new technologies would be viewed by other countries in the context of their national goals. This research effort would develop the tools and approaches for understanding, limiting, and managing the risks of nation-state proliferation and physical security for fuel cycle options. NE will focus on assessments required to inform domestic fuel cycle technology and system option development. These analyses would complement those assessments performed by NNSA to evaluate nation-state proliferation and the international nonproliferation regime. Taken in conjunction, these comprehensive proliferation risk assessments will provide important information for discussions and decisions regarding fuel cycle options. These assessments will:
  - Exploit science-based approaches, to the extent possible, for analyzing difficult-to-quantify proliferation risk factors or indicators (e.g., capabilities, motivations, and intentions); address issues identified in several National Academy of Sciences studies related to risk assessment; and leverage current state-of-the-art academic social science research in this field.
  - Integrate the diverse decision factors (including economics, public health and safety, environmental benefits, and proliferation and terrorism risk reduction) for different fuel cycle options to

understand the tradeoffs and potential synergies between these decision criteria.

- Apply these tools to study nuclear energy system options, and display the results in a useful format for decision makers.

- *Safeguards and Physical Security Technologies and Systems* – The NE focus is on the development of safeguards technologies and integrated systems for current and potential future domestic fuel cycle options. These technologies and systems contribute significantly to limiting proliferation and physical security risks while also achieving economic, public health and safety, and environmental goals. This requires that these activities be performed in an integrated program with the fuel cycle technology development activities. Opportunities exist to collaborate with other organizations (e.g. NNSA, the Department of Homeland Security, the Department of Defense) and will be utilized. NNSA will be responsible for evaluating the nation-state proliferation risks of deploying new fuel cycle technologies – particularly recycling technologies – outside of the United States.
  - *Advanced Instrumentation* – Many advanced fuel cycle processes, such as advanced aqueous reprocessing, electrochemical separations, and recycle fuel fabrication pose new challenges for safeguards and nuclear material management. The safeguards state- of-the-art will be advanced through a developmental program to improve the precision, speed, sampling methods, and scope of nuclear process monitoring and accountancy measurements, and innovative approaches for containment and surveillance. This effort supports the development of advanced safeguards instrumentation such as active interrogation methods based on neutron and photon drivers and advanced passive detectors, such as ultra-high resolution spectrometer and neutron multiplicity counting. Additionally, existing nuclear data is evaluated for the identification of gaps or needed improvements.
  - *Advanced Concepts and Integration* – Early integration of safeguards concepts into nuclear facility design is optimal to meet U.S. and international standards with minimum impact on operations. This requires development of a framework to codify the safeguards-by-design concept, applicable for both international safeguards and physical security for U.S. fuel cycle facilities. It also includes the evaluation of material attractiveness of relevant fuel cycle materials. A monitoring and control system

must be developed that is secure and can rapidly authenticate and investigate summary and raw data to unequivocally distinguish process deviations, maintenance problems, and calibration and component failures from actual diversion events.

  - *Modeling and Simulation* – Development of modeling and simulation tools to enable new technology development, elucidation of high-impact R&D priorities, and approaches that optimize effectiveness and efficiency of the overall system will be essential for the integration of new safeguards technologies and techniques into nuclear energy systems.

- *Nuclear Energy Technologies and Systems* – This element includes developing and assessing a sufficiently wide and innovative range of options (in concert with R&D Objectives 1–3) to achieve Objective 4. This includes, for example, options that enable decreasing the attractiveness and accessibility of used fuel and intermediate materials, transmuting materials of potential concern, optimizing safeguards and physical security systems approaches, and minimizing the number of needed enrichment and recycle facilities. In conjunction with NNSA, NE will lead the development of these options and implement mechanisms that tightly link and inform both this R&D and other elements of R&D Objective 4.

4.4.3. Key Activities and Milestones

The following chart outlines potential milestones and future national industry aims for this objective. It presents a distinction between near-term milestones toward which the NE R&D plan is designed to progress, represented as triangles, and longer-term potential outcomes that provide a framework for the milestones, shown as ovals. The milestone charts attempt to depict the stages of development so as not to leave a sense that new technologies can be immediately deployed at a commercial level. Not every milestone or potential outcome outlined in these charts represents actions that are within DOE's roles and responsibilities, and research paths will include many decision points that require choosing the most promising options for continued R&D. All DOE R&D activities will be evaluated and revisited regularly and modified as necessary through the budget process to ensure the portfolio reflects past progress and current priorities.

Although some smaller component or process "demonstration" activities are mentioned, these are largely field tests and other actions to provide proof or validation of system elements. They are not costly, large-scale demonstrations like NGNP. Any consideration to embark on such large-scale demonstrations will be the result of decision-making and budget development processes.

5. R&D Approach

Section 4 of this roadmap presents NE's four R&D objectives. These objectives show the connection between how nuclear energy will contribute to meeting the nation's energy goals and the R&D that needs to be performed to enable that contribution. This section describes the approach that will be taken to perform this R&D, provides brief descriptions of the key areas of technological development that will be undertaken, presents a brief description of the facilities needed to perform this research, and describes the interfaces with stakeholders that will be required for success.

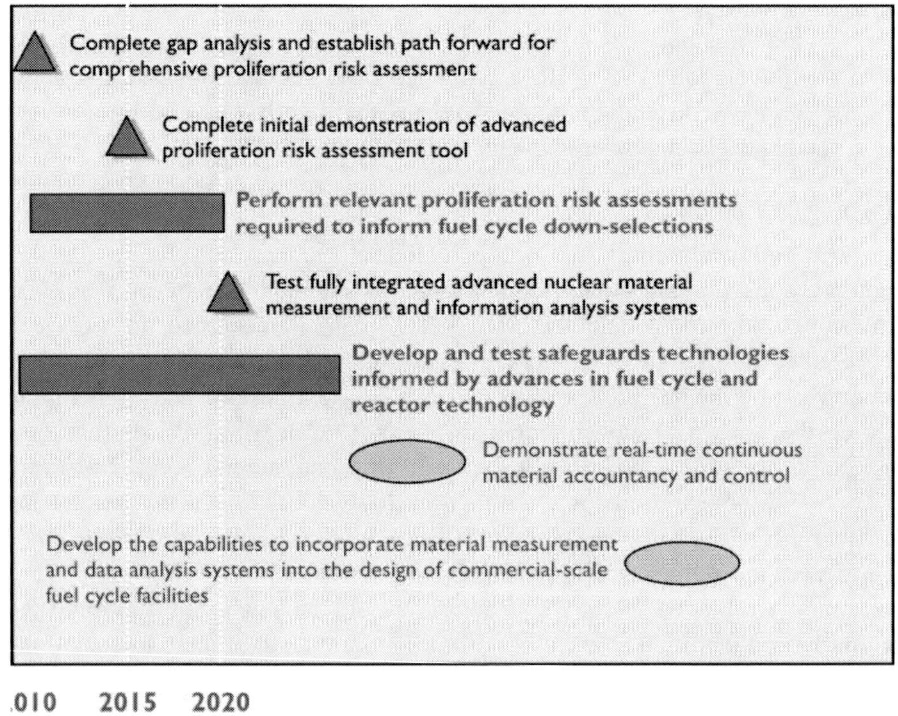

Figure 13. Key Activities for R&D Objective 4.

5.1. Solution-Driven, Goal-Oriented, Science-Based Approach to Nuclear Energy Development

Nuclear power systems were initially developed during the latter half of the 20th century. Their development was greatly facilitated by the nation's ability and willingness to conduct large-scale experiments. The federal government constructed 52 reactors at what is now Idaho National Laboratory, another 14 at Oak Ridge National Laboratory, and a few more at other national laboratory sites. By today's standards, even large experiments and technology demonstrations were relatively affordable. While relying heavily on the Edisonian approach in the 1950s and 1960s, the nuclear energy community was a rapid adopter of high-end computational modeling and simulation during the 1970s and 1980s. During this period, nuclear power plant designers and regulators developed and deployed many of the most demanding simulation models and tools on the most advanced computational platforms then available. Still, the United States embraced a regulatory process that relied, and still relies, heavily on the use of experiments to confirm the ultimate safety of nuclear power systems. Building upon the scientific advances of the last two decades, our understanding of fundamental nuclear science, improvements in computational platforms, and other tools can now enable a new generation of nuclear power plant designers, fabricators, regulators, and operators to develop technological advancements with less of a reliance on large-scale experimentation. The developmental approach employed in this roadmap embodies four elements, as depicted in Figure 14:

Experiments – These are generally small-scale experiments aimed at observation of isolated phenomena or measurements of fundamental properties. However, targeted integral experiments also will be needed in some cases.

Theory – Based either on first principles or observations made during phenomenological testing, theories are developed to explain fundamental physical phenomena.

Modeling and Simulation – A range of mathematical models for diverse phenomena at much different time and spatial scales are developed and then integrated to predict the overall behavior of the system. Key objectives of the modeling and simulation effort are to reduce the number of prototypes and large- scale experiments needed before demonstration and deployment and to quantify uncertainties and design and operational parameters.

Demonstrations – While the state of knowledge can be significantly advanced through the combination of experiments, theory, and modeling and simulation, there may be instances where it is appropriate to work with the private sector to further develop and validate laboratory findings. Demonstrations can be a useful element in proving viability of new technologies, but their high cost must be considered in the context of a variety of other factors. There must be sufficient industry commitment for deployment of commercial technologies before such demonstrations would be considered. Any potential future demonstration activities will be evaluated on a caseby-case basis through the established decision-making procedures of the Department and budget formulation.

5.2. Enabling Technologies

A set of enabling technologies has been identified that support progress on multiple objectives. Where NE has an R&D role in these technology areas, coordination of NE's activities across these technologies must be implemented. For example, the NE "owner" of the fuel cycle objective in such a case will be responsible for coordination of all nuclear fuel work across objectives.

- *Structural Materials* – Advanced radiation and corrosion-resistant materials with extension to high-temperature applications benefit many of the R&D objectives, especially when conducted using a science-based development approach without relying heavily on empirical experiments. Thus, a synergistic R&D program can be developed to support all the objectives.
- *Nuclear Fuels* – The development of improved and advanced nuclear fuels is clearly a major objective for both existing LWRs and the entire spectrum of advanced nuclear energy systems discussed throughout this document. The short list of potentially needed fuels include high-burnup LWR, fast reactor, and gas-reactor fuels; coated-particle fuels; fast-spectrum and thermal-spectrum transmutation fuels and targets; thorium fuels; and molten-salt fuels. A tightly coordinated

and well integrated nuclear fuels R&D program must be developed to support all of the R&D objectives.

- *Reactor Systems* – The development of advanced reactor concepts and supporting technologies is a core function of NE. Advanced technologies and reactor concepts are needed to improve the economics of electricity production. Multiple advanced reactor concepts (LWR, small modular, gas-cooled, liquid metal-cooled, molten salt-cooled, etc.) may play a role in our nuclear future. The development of a robust advanced reactor system concept definition capability will be an important element of NE strategy development.
- *Instrumentation and Control* – The development and implementation of digital instrumentation and control systems will benefit current reactors as well as future reactors. Advanced instrumentation and control systems will also benefit future fuel cycle facilities. Safeguards technology development also relies on advanced instrumentation and plant control systems through safeguards-by-design.

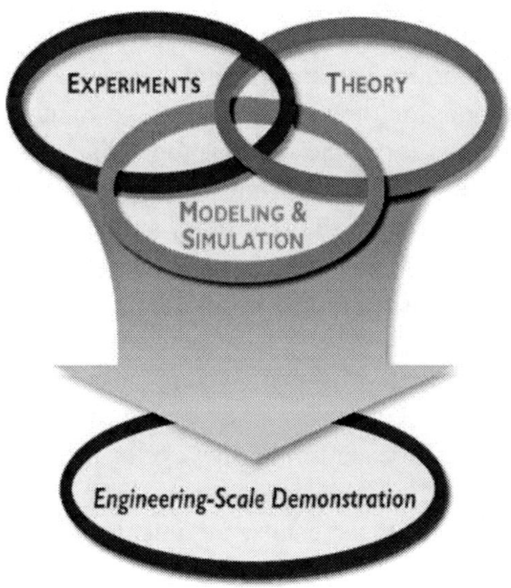

Figure 14. Major Elements of Science-Based Research, Development & Demonstration.

- *Power Conversion Systems* – Advanced power conversion systems will lead to increased efficiency for the future reactors and facilitate the use of nuclear power in markets requiring process heat.
- *Process Heat Transport Systems* – The development of process heat transport systems that can be combined with multiple reactor technologies will enable the use of nuclear power to deliver needed process heat to the industrial sector.
- *Dry-Heat-Rejection Systems* – Advanced dry-heat-rejection systems will improve the environmental friendliness of the nuclear power plants and enable the deployment of nuclear energy in areas where water constraints might otherwise preclude its use.
- *Separations Processes* – This report has noted the wide variety of fuel cycle options that may be needed in the future to address U.S. energy security, economic, and sustainability goals. Our future ability to sustainably and economically recycle LWR fuels, fast reactor fuels, gas-cooled reactor fuels, molten salt fuels, *etc.* will depend, in part, on our ability to separate key elements from the waste that will not be disposed in a repository.
- *Waste Forms* – The ability to engineer, produce, and manage fuel cycle waste forms that are chemically and structurally stable over relevant periods of time from decades to hundreds of thousands of years (depending on the radioisotope) is critical to achieving a sustainable fuel cycle and must be closely integrated with both radiochemical research and repository systems research.
- *Risk Assessment Methods* – Advanced methods for risk assessment based on mechanistic modeling of system behavior will benefit the safety assessments of the new nuclear energy systems and fuel cycle technologies. State-of-the-art computational and experimental techniques will benefit not only novel reactor concepts but other nuclear facilities needed for the fuel cycle.
- *Advanced Modeling and Simulation* – The science-based approach relies heavily on fundamental experiments combined with associated theories for predictive capabilities. However, a comprehensive use of the science-based approach for predictive tools with multiple interrelated phenomenologies requires advances in computational sciences where phenomena at different time and length scales can be bridged into an engineering code using modern computational platforms.

5.3. R&D Facilities and Infrastructure

Ultimately all design and safety tools for nuclear systems must be validated with underpinning experimental data. Without such a foundation in reality, licensing these systems would be virtually impossible. Experiments also provide essential waypoints for guiding the development of technology. Having such an experimental capability requires that nuclear energy R&D maintain access to a broad range of facilities from small-scale laboratories potentially up to full prototype demonstrations. Hot cells and test reactors are at the top end of the hierarchy, followed by smaller-scale radiological facilities, specialty engineering facilities, and non- radiological small laboratories.

Nuclear energy R&D employs a multi-pronged approach to having these capabilities available when needed. The core capabilities rely on DOE-owned irradiation, examination, chemical processing and waste form development facilities. These are supplemented by university capabilities ranging from research reactors to materials science laboratories. Future infrastructure requirements will be considered through the established budget development processes as needs arise.

The high cost of creating and maintaining physical infrastructure for nuclear R&D, including the necessary safety and security infrastructure, requires creativity and periodic realignment of infrastructure planning with programmatic direction. NE successfully employs a solid approach to maintaining infrastructure. The approach concentrates the high-risk nuclear facilities at the remote Idaho site, maintains unique capabilities at other sites if required, supports vital university infrastructure, negotiates equitable capability exchanges with trusted international partners, refurbishes and reequips essential facilities if required, addresses maintenance backlogs to ensure safe operation, and makes efficient use of modeling, simulation, and single-effect experiments.

5.4. Interfaces and Coordination

In order to achieve the objectives under each R&D objective, NE must closely coordinate its activities with other agencies, the nuclear industry, and international partners.

Other Department of Energy Offices –The use of a "science-based" approach to develop innovative nuclear energy systems and components requires a strong collaboration between NE and the Office of Science (SC) to employ the tools developed for science in engineering applications. Such tools include advanced experimental techniques, a fundamental understanding of materials behavior, and advanced computational sciences. R&D on storage and disposal of nuclear waste will be performed in coordination with the Office of Environmental Management (EM) and the Office of Naval Reactors (NR), as there are salient similarities in the disposition challenges facing each.

NNSA – Technology development for safeguards is a crosscutting tool that is applicable for both domestic and international uses. NNSA and NE are implementing a coordinated effort to address the safeguards R&D needs for domestic and international applications. These collaborative efforts address the assessment of proliferation risks, accountancy, and control (domestic) and verification (international) by contributing new safeguards technologies; recruiting a new generation of safeguards specialists into the U.S. national laboratories, universities, and industry; and informing the development of safe and secure nuclear facilities. NNSA will be responsible for evaluating the international nation state proliferation risks of deploying new fuel cycle technologies, particularly recycling technologies, outside of the United States.

NRC – Appropriate collaboration between DOE and the NRC will help assure that nuclear energy remains a viable option for the United States. The development of science-based tools to inform licensing paradigms is one key goal of this collaboration.

Nuclear Industry – The decision to deploy nuclear energy systems is made by industry and the private sector in market-based economies. However, it is important that industry is engaged during the definition and execution of the R&D phase and that industry participate in joint demonstration activities if such demonstration is deemed necessary and appropriate to facilitate commercialization and deployment of the resulting technologies and systems. As technologies are developed, cost-sharing with industry is an integral part of NE's agenda. DOE will proceed in a manner that recognizes the importance of maintaining a strong and viable nuclear industry.

International Community –Strong participation and leadership by the United States in international nuclear R&D, safety and nonproliferation programs is essential. Nuclear energy worldwide must be deployed with safety and security of paramount importance. In addition, several countries have established strong nuclear R&D programs and specialized expertise from which the United States can benefit, such as the leadership position of Russia, France, and Japan in fast reactor technology. Collaborations in nuclear technology R&D will be implemented through bilateral and multilateral agreements and through international organizations such as the Generation IV International Forum.

In order for nuclear power to continue to be a viable energy option in any country, including the United States, nuclear safety, security, and safeguards must be maintained at the highest levels on a global scale. DOE will help to achieve consensus criteria for safe reactor operation through international organizations, such as the World Association of Nuclear Operators, and seek to enhance safety standards for nuclear power, promote appropriate infrastructure at the national and international levels, and minimize proliferation risks from the expansion of nuclear power through its participation with the IAEA and related organizations.

6. Summary and Conclusions

This document presents an integrated strategy and R&D framework for the DOE Office of Nuclear Energy. In order to meet the Administration's goals of energy security and greenhouse gas reductions, nuclear energy must play an important role in the national energy portfolio. NE's derived missions in support of these national goals are to enable the development and deployment of fission power systems for the production of electricity and process heat. Four research and development objectives have been identified, which will guide NE's program and strategic planning. Progress in these areas will help ensure that nuclear energy continues to be among the suite of available U.S. energy options throughout the 21st century. These objectives are:

- *R&D Objective 1* – Develop technology and other solutions that can improve the reliability, sustain the safety, and extend the life of current reactors.
- *R&D Objective 2* – Develop improvements in the affordability of new reactors to enable nuclear energy to help meet the Administration's energy security and climate change goals.

- *R&D Objective 3* – Develop sustainable nuclear fuel cycles.
- *R&D Objective 4* – Understand and minimize the risks of nuclear proliferation and terrorism.

End Notes

[1] 2007 GHG emissions reported in EPA, *Inventory of U.S. Greenhouse Gas Emissions and Sinks: 1990 – 2007* EPA 430-R-09-004, April 15, 2009. Administration emission goals taken from the "Testimony of Peter R. Orszag, Director of the Office of Management and Budget, Before the Committee on the Budget, U.S. House of Representatives" on March 3, 2009.

[2] The index was developed by the United Nations to enable cross-national comparisons of development and is updated in an annual report. The derivation of the index was introduced in United Nations Development Programme, *Human Development Report 1990*, Oxford University Press, 1990.

[3] The data in Figures 5 and 6 are reported by the U.S. DOE Energy Information Agency "An Updated Annual Energy Outlook 2009 Reference Case," 2009.

[4] R. Socolow and S. Pacala, "Stabilization Wedges: Solving the Climate Problem for the Next 50 Years with Current Technologies." *Science*, August 13, 2004: 968-972.

[5] *Scientific American*, September 2006

[6] EIA, *Annual Energy Review* 2008, Table 9.2.

[7] Ref. http://www.gallup.com/poll/117025/Support-Nuclear-Energy-Inches-New-High.aspx.

In: Nuclear Energy
Editor: Petr Makrygina
ISBN: 978-1-61470-999-2

Chapter 2

NUCLEAR ENERGY ENABLING TECHNOLOGIES WORKSHOP REPORT*

United States Department of Energy

ACRONYMS AND INITIALISMS

ASME	American Society of Mechanical Engineers
BIM	Building Information Modeling
BWR	Boiling Water Reactor
CAD	Computer-Aided Design
CAM	Computer-Aided Manufacturing
CBM	Condition-Based Monitoring
CCF	Common-Cause Failure
CMOS	Complementary Metal-Oxide Semiconductor
CO_2	Carbon Dioxide
Cr	Chromium
DOE	Department of Energy
DPA	Displacements Per Atom
EDXRD	Energy Dispersive X-Ray Diffraction
FEA	Finite Element Analysis
FM	Ferritic–Martensitic

* This is an edited, reformatted and augmented version of a United States Department of Energy publication, dated July 29, 2010.

FY	Fiscal Year
GEN	Generation
GMAW	Gas Metal Arc Welding
GTAW	Gas Tungsten Arc Welding
GWD/MTU	Gigawatt-Days per Metric Ton Uranium
HT-UPS	High-Temperature–Ultrafine-Precipitation-Strengthened
HSI	Human–System Interface
IAEA	International Atomic Energy Agency
I&C	Instrumentation and Controls
INPRO	International Project on Innovative Nuclear Reactors and Fuel Cycles
ITAAC	Inspections, Tests, Analyses, and Acceptance Criteria
ITER	International Thermonuclear Experimental Reactor
KSI	Kilopounds per Square Inch
LIBS	Laser-Induced Breakdown Spectroscopy
LWR	Light Water Reactor
LWRS	Light Water Reactor Sustainability
MPACT	Materials Protection, Accounting, and Control Technology
Na	Sodium
NASA	National Aeronautics and Space Administration
Nb	Niobium
NDE	Non-Destructive Examination
NE	Office of Nuclear Energy
NEET	Nuclear Energy Enabling Technologies
NGNP	Next Generation Nuclear Plant
NNSA	National Nuclear Security Administration
NRC	Nuclear Regulatory Commission
ODS	Oxide Dispersion- Strengthened
PDF	Pair Distribution Function
PID	Proportional–Integral–Derivative
PIE	Post-Irradiation Experiment
PP&PP	Physical Protection and Proliferation Resistance
PRA	Probabilistic Risk Assessment
PRD	Proposed Research and Development
PWR	Pressurized Water Reactor
R&D	Research and Development
RFID	Radio Frequency Identification

RIA	Reactivity-Initiated Accident
RPV	Reactor Pressure Vessel
SAPRA	Simplified Approach for the Proliferation Resistance Assessment of Nuclear Systems
SAXS	Small-Angle X-Ray Scattering
SCC	Stress Corrosion Cracking
SiC	Silicon Carbide
SMR	Small Modular Reactor
TEM	Transmission Electron Microscopy
WAXS	Wide-Angle X-Ray Scattering

1. INTRODUCTION

The Nuclear Energy Enabling Technologies (NEET) Program proposed for fiscal year (FY) 2011 at the requested $99 million will develop crosscutting technologies that directly support and complement the Department of Energy, Office of Nuclear Energy's (DOE–NE's) advanced reactor and fuel cycle concepts, focusing on innovative research that offers the promise of dramatically improved performance. Pending FY 2011 Congressional Appropriation, the program will encourage the development of transformative, "out-of-the-box" solutions across the full range of nuclear energy technology, spurring revolutionary improvements in safety, performance, reliability, economics, and proliferation risk reduction. The NEET Program consists of three elements, the first two of which are the subjects of this report:

- Crosscutting Technology Development
- Transformative Nuclear Energy Concepts Research and Development (R&D)
- Energy Innovation Hub for Modeling and Simulation

Crosscutting Technologies to Meet R&D Objectives

DOE–NE will provide R&D support for existing and future nuclear energy concepts in areas such as reactor materials, advanced methods for manufacturing and construction, new sensor technologies and instrumentation, and creative approaches to understand and reduce proliferation risks and terrorism.

Transformative Nuclear Concepts R&D

This program element supports high-risk–high-reward concepts that have the potential for making significant leaps forward in advanced nuclear technology development. Research on transformative nuclear concepts will pursue non-traditional nuclear energy ideas that offer the potential for improved system performance and may radically alter nuclear system configuration and development needs. Possible R&D topics include development of specialized nuclear fuels, revolutionary materials, coolants, new techniques for energy conversion, waste disposal, nonproliferation, and other innovations. DOE–NE will support an open competitive solicitation process for transformative investigator-initiated projects.

1.1. Key Challenges Facing Nuclear Energy

According to the *Nuclear Energy R&D Roadmap Report to Congress*, submitted in April 2010, the key challenges facing the nuclear energy industry are:

- Capital Cost – Develop innovative designs to reduce capital costs; smaller reactors, modular construction.
- Waste Management – Transition to nuclear energy technologies that significantly reduce the production of long-lived radioactive waste.
- Proliferation and Terrorism Risks – Develop and demonstrate options that limit proliferation and terrorism risks while also achieving economic, public health and safety, and environmental goals.
- Safety and Reliability – Maintain an excellent safety and reliability record of nuclear energy in the United States and share relevant technologies with other countries.

The roadmap takes a science-based approach to resolve these challenges that includes a combination of experiments, theory, modeling and simulation, and demonstration.

1.2. DOE Nuclear R&D Objectives

The roadmap further enumerated DOE's nuclear R&D objectives:

- Develop technologies and other solutions that can improve the reliability, sustain the safety, and extend the life of current reactors.

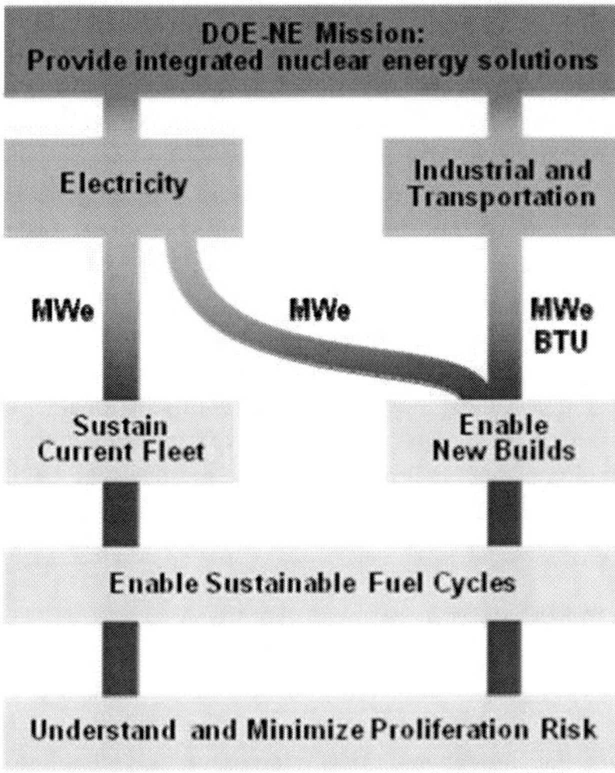

Figure 1. The DOE–NE mission.

- Develop improvements in the affordability of new reactors to enable nuclear energy to help meet the Administration's energy security and climate change goals.
- Develop sustainable nuclear fuel cycles.
- Understand and minimize the risks of nuclear proliferation and terrorism.

These objectives will serve as guidelines to develop the NEET Program. NE has outlined some of the specific issues facing the nuclear industry in each of these four areas. Challenges to extending plant lifetime include aging and degradation of system structures and components, fuel reliability and performance, obsolete analog instrumentation and control (I&C) technologies, and design and safety analysis tools. Challenges facing new nuclear reactor plants include financial hurdles, developing small reactors to reduce up-front capital costs, developing plant designs that address industrial needs, and uncertainty over new regulatory frameworks. Challenges to achieving a sustainable fuel cycle include developing high burn-up fuel and structural materials to withstand irradiation for longer periods of time, developing simplified separations and waste management, and developing optimized systems to maximize energy production while minimizing waste. Challenges associated with minimizing proliferation and terrorism risks include the potential for misuse of technology and materials for weapons proliferation purposes, theft, or sabotage. These challenges will be addressed through innovative nuclear reactor and fuel cycle technologies and systems; science-based next-generation materials protection, accounting, and control technologies and systems; support for new international framework and institution development; and new approaches for understanding and managing risks for technologies and integrated fuel cycle options. The DOE-NE mission is highlighted in Figure 1.

1.3. NEET Program Workshop Objectives

DOE's primary purposes in convening the NEET workshop were to:

- Obtain stakeholder input on the crosscutting technology R&D needs to support NE's R&D roadmap objectives.
- Inform a possible solicitation for transformative, "out-of-the-box" solutions across the full range of nuclear energy technology issues.

Pending FY 2011 Congressional Appropriation, DOE–NE will solicit nuclear industry stakeholders to propose ideas on crosscutting and transformative nuclear technologies and capabilities for incorporation into advanced reactor and fuel cycle concepts that offer the promise of revolutionary improvements in safety, performance, reliability, economics, and minimization of proliferation and terrorism risks; and promote creative solutions to the broad array of nuclear energy problems related to reactor and fuel cycle development. Researchers are encouraged to identify barriers towards enacting those technologies and find ways to surmount those barriers. The innovative type of research promoted through the NEET program will support technical leaps forward in nuclear technology. Transformative Nuclear Concepts R&D is not meant to fund conservative R&D that achieves merely incremental progress, but high-risk–high-reward concepts that truly push beyond currently conceived research.

NEET has key relationships to other long-term DOE initiatives as identified in the FY 2011 budget request: the Small Modular Reactors (SMR), Next Generation Nuclear Plant (NGNP), and Light Water Reactor Sustainability (LWRS) programs. National laboratories, universities, and industry R&D communities will be engaged to enable SMR designs. This effort entails R&D in materials, fuels, I&C, and fabrication research and testing; modeling and simulation of reactor systems and components; and probabilistic risk analyses of innovative design features and safety systems. An emphasis will be placed on simplified operation and maintenance, enhanced functionality, and increased proliferation resistance and security. Moving the NGNP forward will require fuel development and qualification, establishment of a licensing basis for gas-cooled reactor coated-particle fuel, graphite materials qualification, and high-temperature materials qualification. It will be necessary to develop and benchmark improved simulation techniques as well as analytical codes and methods. To sustain the existing light water reactor (LWR) fleet requires R&D to acquire science-based fundamental understanding of materials aging and degradation and to develop long-life fuel using advanced materials. Advanced instrumentation, information, and control system technologies also need to be developed along with improved human–machine interface capabilities.

The workshop had Crosscutting Technology Development breakout sessions and a Transformative Nuclear Concepts R&D session.

During the Crosscutting Technology Development Breakout sessions, the invited panel members provided presentations on the "current state-of-the-art," "future state," "knowledge gaps," and "R&D needs" for their specific technology areas. The audience participated in the discussions and provided input to these areas. This report captured the ideas presented and discussion during the breakout session. These do not necessarily represent a complete set of research needs in these areas.

During the Transformative Nuclear Concepts R&D session, the invited panel members provided presentations on the "current state-of-the-art," "grand challenges to be solved," and "proposed R&D to solve the grand challenges" for their specific technology areas. This report captured the ideas presented and discussion during this session. These were examples of some of the technologies areas; the Transformative Nuclear Concepts R&D planned solicitation, pending FY 2011 Congressional Appropriation, will not be limited to these examples. All high-risk-highreward concepts that have the potential for making significant leaps forward in advanced nuclear technology development will be considered.

2. Discussion of Crosscutting Technologies

The NEET Program workshop was organized into technology-centered sections:

- Methods for Manufacturing and Construction
- Proliferation Risk Assessment
- Reactor Materials
- Advanced Sensors and Instrumentation

Presenters in each session were encouraged to organize and analyze crosscutting concepts applicable to more than one technological area or more than one research program. For each technology, panelists described the research activity and its relevance to DOE–NE programs, the current state-of-the-art, and a desired future state, along with the gaps in technology and proposed R&D to achieve that future state. Panelists were encouraged to:

- Delineate the proposed technical activities and related elements in other DOE-sponsored R&D programs, considering near- and long-term research.
- Discuss gaps between needed and existing capabilities that may be addressed through this R&D program.
- Identify areas for coordinated research between this program and other R&D efforts, considering possible cost-sharing with the private sector and university interest in related educational programs.

2.1 Advanced Methods for Manufacturing and Construction

Since the 1980s, U.S. industry's manufacturing and construction activity in the nuclear energy sector has been very limited—primarily aimed at maintenance, retrofits, and upgrade of existing nuclear plants in the United States and support of construction projects overseas. In the interim, plant designers have developed new nuclear plant designs that will depend upon new and advanced methods and technologies for manufacturing and construction to improve quality while reducing costs and schedules. These new standardized designs, referred to as Generation III or III+ (Gen III+), have either already been approved by the U.S. Nuclear Regulatory Commission (NRC) through design certification or are currently in the NRC certification process.

Some of the new standard plant designs are being built overseas before being used in U.S. construction projects. The U.S. projects will commence on several sites over the next few years. This overseas head-start provides an opportunity to obtain and apply lessons learned from the overseas projects. However, oversees experience also points out the importance of quickly developing and applying new manufacturing, fabrication, and construction technologies in the United States that are competitive in a global marketplace, to reduce the costly new plants, and help to re-establish U.S. technological leadership.

In addition to the standard plant designs currently in the NRC certification process, even more advanced designs, e.g. the Small Modular Reactors (SMR), will soon be submitted for NRC certification. These new designs will rely on even more advanced technologies for fabrication and construction—e.g., much more extensive use of modularization and factory fabrication.

This section on advanced manufacturing and construction contains ideas discussed at and after the workshop. It proposes innovative methods for fabrication and component manufacturing that will help reduce the cost and schedule of new nuclear plant construction. Fabrication innovation may include laser welding processes, integrated welding and non-destructive examination (NDE) systems, modular pipe fabrication/welding, ultrasonic machining, modular wiring and cabling systems, fast and reliable on-site material characterization systems, and new pipe/component methods for manufacturing. Construction innovation will enhance modular building using composite structures, seismic base isolation systems, pre-assembled rebar systems, new rebar materials; transport, lifting, and rigging improvements; and modeling and simulation to improve design, and construction sequencing.

2.1.1. Overview/Current State of the Art

Global construction and manufacturing methods provide the U.S. with the vision to go from a "stick built" construction process of the past to a future of innovation in new U.S. power plant construction development. The U.S. will build with new technologies in all areas of the Engineering, Procurement and Construction (EPC) process. *Engineering* is defined as the design elements of a new build, *procurement* as the products and services that must be manufactured and fabricated to support the *construction* of the plant. EPC was used to focus the workshop information.

- Engineering
  - Advances in modeling and simulation
  - Design codes for Steel Plate Concrete Composite Construction
  - Base isolation systems for seismic systems
- Manufacturing and fabrication (Procured items)
  - Welding and inspection technologies
  - Near net shape manufacturing using powder metal or additive manufacturing
  - Ultrasonic machining
  - Hybrid laser systems
  - Modular systems
- Construction
  - Modular rebar systems
  - New rebar materials
  - Advances in concrete formulations
  - Models for fabrication, modularization, and construction sequencing

In all of these areas the current practices used in Japan, Korea, France, and other countries that have continued to develop their EPC for nuclear plant construction were explored along with innovation in the U.S. and other countries regarding manufacturing technologies, particularly those used in other heavy industries like ship building, petro-chem and aircraft construction. The workshop and subsequent discussions demonstrate that the "art of the possible" can be expanded to redefine how the U.S. builds new power plants.

2.1.2. Future State

To be successful in the future, industry EPC standards will have to adopt pre-planning approaches, first-of-a-kind situations, and readiness reviews in planning. Industry must fully recognize the value of effective planning of new nuclear projects and support projects with adequate cash flow and completion of engineering to the greatest extent possible prior to the start of construction.

Engineering

The design development process will use 3-D modeling as the basis for all work. Computer- Aided Design/Computer-Aided Manufacturing (CAD/CAM) and Building Information Management (BIM) systems will be integrated with all the engineering analysis and manufacturing elements to provide a common platform for the EPC and operation of the plant.

Designs using base seismic isolation systems, particularly in SMR's, will be used to reduce, minimize, or eliminate the unique requirements of site specific seismic analysis. This can greatly reduce the numbers of supports and snubbers and well as reduce the severity of seismic loads imposed on equipment thus helping to reduce cost.

Designs and validated models for concrete composite systems will improve the ease of construction and costs of containment and large civil structures. It will also provide the regulatory authorities with a method to evaluate the design of these structural systems.

Manufacturing (Procured Items)

Future nuclear plants will utilize high-productivity welding, ultrasonic welding, and integrated welding and inspection technologies. The nuclear industry does not fully utilize recent fabrication advances, especially laser hybrid technology. Despite dramatic improvements, this technology still faces hurdles. Historically, laser hybrid welding has been expensive and cumbersome. However, the advent of fiber lasers in recent years has reduced cost and improved efficiency. These new lasers offer the efficiency and portability necessary for their broad implementation and acceptance in the nuclear industry. In the future, nuclear plants will be built utilizing high-productivity welding and integrated welding and inspection technologies. These methods will improve quality, reliability and cost of the components and systems on a power plant. Components will be manufactured using new metals and metal working methods allowing production of near-net-shape sub assemblies, such as valve and pump bodies, that significantly reduce cost and manufacturing schedules.

Ultrasonic machining will produce low machine forces, increased portability, and improved tool life, with a simultaneous increase in productivity. For example, the final machining of a typical reactor closure head requires three to four weeks of non-stop machining to complete; the integration of ultrasonic machining into the same machining cell, using existing machining centers, would reduce the machining time by half.

Future devices enabling integrated welding will take into account environmental conditions such as temperature, humidity, real-time microstructure monitoring/prediction, close proximity NDE (within a few inches of the welding operations), and local global positioning systems (GPS). By accurately monitoring welding torch position and orientation, environmental conditions, microstructure and weld cooling rates, and pseudo real-time NDE, the welding systems of the future will be able to achieve true closed-loop weld quality control. Figure 2 shows an example of the latest in laser hybrid welding technology.

Figure 2. The latest in laser hybrid welding technology, a portable device that does not require CO_2 tubes.

Construction

Future plants will likely make extensive use of steel–concrete composite structures—a technology that has the potential of significantly reducing the cost and duration of building large civil structures such as containment and other safety-related buildings. In the future, industry will make full use of the new cranes currently being designed and eventually lead the way to the creation of super-large crane designs for installation of larger prefabricated modules to create modular components, engineers and contractors will make extensive use of automated manufacturing with direct CAD/CAM linkage.

The essential elements of civil construction, concrete and steel, will be redeveloped using improved concrete compositions that are more uniform and have more predictable physical properties. Rebar (reinforcing bar) will be higher strength and more corrosion resistant. This will allow lower rebar densities and better concrete placement, thus avoiding concrete deficiencies like voiding and honeycomb. An example of reduction in rebar density is shown in Figure 3. Rebar will also be placed as modules, not as individual pieces. The rebar modules will be built in factories or fabrication assembly buildings and will be based on the models discussed under the Engineering sections.

Figure 3. Reduction in rebar density with the use of high-strength rebar (conventional 60 ksi rebar [left] versus 100 ksi high-strength rebar [right]).

Future plants will not be constructed in the same way they were in the past. Manufacturing, fabrication, and construction methods will focus on increasing levels of modularization, which reduces construction duration and reduces financing costs—two of the biggest factors in the assuring feasibility of future new nuclear plant deployment.

2.1.3. Knowledge Gaps

To get to the future state, there are many knowledge gaps that must first be closed. In general, there needs to be EPC processes that are modeled, manufactured and constructed using a platform that supports the new build technologies, like modular construction, but also can be maintained and used for the life of the plant. The construction and manufacturing processes used around the globe must be translated to our new construction activities. More needs to be built in the shop using methods that reduce cost and improve reliability. Specific improvements are needed in the following areas:

- Engineering design software and modeling tools
- Methods of manufacturing to improve cost, delivery and reliability
- Civil construction methods which reduce the "stick built" designs of the past and take advantage of new methods for concrete and steel placement.
- Mitigation of dynamic seismic hazards, possibly utilizing base isolation technologies. Other gaps prevent U.S. manufacturers from competing with the ever-growing low-cost international market. Heavy component manufacturing needs to incorporate:
- Increased automation
- Adapted regulations to allow for more on-machine inspections
- Innovative approaches to large component /manufacturing
- Improved weld, inspect, and repair processes
- Innovation in component manufacturing and process metallurgy

Improvements do not stop in the factory. Additional development is needed in setting requirements to maintain design and quality certifications through transportation to the site.

Lessons learned from currently ongoing fabrication and construction activities supporting overseas nuclear plant projects must be collected and applied to U.S. projects.

For some portions of the equipment used in the new nuclear plant designs, the competitiveness of U.S. suppliers needs to be determined and methods for improving their competitiveness developed. This should provide opportunities to improve quality and safety while reducing costs and schedule for fabrication/construction.

Where fabrication/construction improvements require regulatory approval (including industry codes and standards), the actions needed, including confirmatory analysis or testing, to obtain regulatory approval need to be considered.

2.1.4. Future R&D

To promote innovative EPC practices, the following are potential areas of investigation:

- Benchmarking and publishing benefits of pre-planning and lessons learned
- Standardizing approaches that reduce reinforcing steel density in nuclear applications

- Encouraging industry adoption of a non-proprietary rebar prefabrication approach that can gain regulatory acceptance, including using mechanical splices for all steel sizes
- Increasing crane capacity while maintaining mobility and increasing lift speeds
- Assisting engineers and contractors with understanding automated manufacturing capabilities and possibilities for CAD/CAM interface
- Developing an integrated suite of BIM–3D CAD–CAM and scheduling/sequencing software that is designed to support the entire plant life cycle. Developing component manufacturing technologies that take advantage of modern materials developments and near-net-shape manufacturing techniques
- New formulations for concrete
- Instrumentation for monitoring concrete defects during or immediately after placement.

To make high-productivity welding widely accepted in the nuclear industry, R&D must address:

- Additional data to justify the return on investment
- Adoption of products into consensus standards
- Regulatory acceptance

Further R&D is necessary to improve integrated welding:

- Future devices will be able to take into account environmental conditions such as temperature and humidity.
- Travel speed will be accurately measured, allowing environmental conditions to be measured.
- When NDE is added to the process, a system will be in place to achieve closed-loop weld quality control.
- Quality algorithms need to be developed, allowing for compilation of all recorded data. Further R&D will also be required to bring advanced component manufacturing to a deployable stage and to gain regulatory acceptance:
- Ultrasonic machining
- Near-net-shape manufacturing using advanced materials and processes

2.2. Proliferation Risk Assessment

Nuclear proliferation and terrorism are important issues as reflected by President Obama's statement: *"We must ensure that terrorists never acquire a nuclear weapon. This is the most immediate and extreme threat to global security"* (Prague, 2009) and the U.S. National Security Strategy, released May 2010: *"There is no greater threat to the American people than weapons of mass destruction, particularly the danger posed by the pursuit of nuclear weapons by violent extremists and their proliferation to additional states."*

Minimizing the risks of nuclear proliferation and terrorism is an integral part of NE's R&D mission and program. This objective includes considering the full spectrum of risks (from both nation-state and sub-national threats), the full range of adversary characteristics (including outsiders, insiders, or combinations of both), and the full spectrum of measures to limit those risks (including innovative nuclear energy technologies and systems; the materials protection, accounting and control technologies that are applied to them; and the frameworks and institutions that create international norms for nonproliferation and security and enable and verify compliance with these norms). A key part of this integrated program is developing new approaches for understanding and managing the risks of nuclear energy technologies and fuel cycle options.

2.2.1. Overview/Current State of the Art

It is important to recognize both the strengths and limitations of risk assessments. Strengths can include the disciplined approach and clear display of important information, including uncertainties, in understandable form so that interested parties can scrutinize key assumptions and, if appropriate, challenge them. Among the most important products of risk assessment are qualitative insights about the structure and performance of complex systems, deeper understanding of dependencies and interactions between different subsystems and components, and fresh, comparative perspectives on the relative advantages and disadvantages of various opportunities to reduce and control risks. Bottom-line absolute risk values, by contrast, should be treated with caution. Risk numbers are not intended to provide the sole—or in many cases, even the principal—basis for decision making. Risk assessment can be helpful, but it constitutes at best only one part of a much more complex process.

For NE, the ultimate goal is to use risk information to advance its mission by effectively informing and guiding its R&D programs, thereby maximizing the prospect of achieving the benefits of nuclear power in a manner that minimizes nuclear proliferation and terrorism risks.

For overall assessments of proliferation and terrorism, multiple methodologies have been developed and applied. These include the Generation IV Proliferation Resistance & Physical Protection (PR&PP) methodology, the International Atomic Energy Agency's (IAEA) International Project on Innovative Nuclear Reactors and Fuel Cycles (INPRO) methodology, and the Simplified Approach for the Proliferation Resistance Assessment of Nuclear Systems (SAPRA). However, while progress has been made, limitations of these approaches, and others, have been identified (for example, see review papers from an earlier related workshop at http://nsspi.tamu.edu/inmm/). A focused R&D program that attacks key issues will advance our ability to understand the relative risks of various potential proliferation pathways and to provide a risk-informed basis for prioritizing investments.

2.2.2. Future State

The state of the art and practice for proliferation and terrorism risk assessments will have progressed and matured so that the tools, applications, and risk management framework are vital and indispensable for proliferation and terrorism risk management and are widely accepted as such. This will enable us to better answer key questions, such as:

- Which innovative concepts and designs will minimize proliferation and terrorism risks?
- How can a nuclear energy technology or system design be optimized to ensure that its operation will minimize the development of skills or technologies that could be used in a nuclear weapons program? (And how can we better maintain assessments as living tools for the lifecycle of technologies and systems?)
- How can the design and operation of a nuclear energy system be optimized to ensure that the system can be safeguarded effectively and efficiently in a safe and economic fashion? Alternatively or additionally, will the optimization delay diversion or misuse?
- What are the benefits and risks of nuclear fuel service concepts?
- How can the global nuclear enterprise architectures be designed to minimize proliferation opportunity?
- On what (and when) do we spend our next dollar to minimize proliferation opportunity?
- What is the risk of the "do nothing" option?

However, important issues to consider as a part of the development efforts include understanding that there is no technological "silver bullet," assuring consistency with and support to U.S. government policy, and understanding and clearly defining questions to be answered and potential utility of the results. Risk management should be the overall objective with a goal of supporting "risk-informed" and not "risk-based" decisions.

Efforts should include soliciting a broad spectrum of stakeholder input, involving experts with risk management experience in other fields, and benchmarking risk management techniques in nuclear power, aviation, the space program, and the Navy. Lessons learned from previous work, including material attractiveness assessments and the importance of broader integrated systems analysis, should be considered and incorporated.

2.2.3. Knowledge Gaps

The Reactor Safety Study (WASH-1400) was the first major application of probabilistic risk assessment (PRA) to reactor safety. The study highlighted gaps in risk analysis including human factors, core melt and containment response, data needs, and common-cause failures (CCFs). In the three decades since the publication of this study, reactor risk assessment has become widely accepted as an essential tool for nuclear safety. The potential value of risk assessment for nonproliferation and security has been recognized, and recent progress has been made in developing methodologies and tools; nonetheless, when comparing the maturity and acceptance of risk assessment and risk management techniques across the two domains of reactor safety and proliferation/terrorism, a large gap remains.

The literature regarding risk assessment for nuclear proliferation and terrorism dates back more than 30 years. In recent years—especially since September 11, 2001—there has been a renewed emphasis on several facets of nuclear-relevant risk analysis work, focused especially on counterterrorism (not limited to nuclear terrorism), homeland security (including not only terrorism but "all hazards"), counterproliferation (which is related to, but distinct from, nonproliferation), and other similar topics.

A key gap highlighted in this work is the ability to understand, analyze, and combat an intelligent, adaptive, and determined adversary. This is an extremely challenging problem, whether or not one uses risk assessment. In addition to the presence of intelligent adversaries, all of the usual challenges of risk assessment apply, including complexity of the phenomena being examined, sparse data, limited models, large uncertainties, difficulties in effectively communicating risk information, and dangers of misinterpretation or misuse. This is partially owing to differences in the problems, but it also provides the impetus for the research topics identified in the Future R&D section below. Additional challenges that have been identified include:

- Change of threats (or perceived threats) with time (days, years, decades)
- Completeness of scenarios being considered
- Impacts of human performance
- Harmonizing design understanding with potential safeguards/ protection possibilities
- Test cases and "benchmark" standards against which these methodologies can be tested (prototypic cases and small systems studies)
- Ability to estimate uncertainties (and sensitivity analysis)
- Expert elicitation in the proliferation/terrorism context
- Lack of empirical data on "failure rates" for nonproliferation and security systems
- Mathematical methods for both technical and behavioral characteristics
- Quantifying the behavior of intelligent adversaries (including terrorists, proliferators, and disgruntled employees)
- Addressing "global market" considerations (requiring dynamic studies with both time and geographic risk transfer)
- Risk metrics that can support comparisons of dissimilar proliferation pathways
- Need for information protection and classification
- Effectively communicating results (to different audiences)

Because the U.S. approach to risk assessment may not be representative of another country's culture, experience and knowledge, analyses that can be both country-specific and country- neutral would be beneficial.

Additionally, there has been confusion regarding terminology, such as the term "proliferation resistance" (the characteristic of a nuclear energy system that impedes one or more states' diversion or undeclared production of nuclear material or misuse of technology in order to acquire nuclear weapons or other nuclear explosive devices). This confusion further challenges development efforts that must include experts from a broad set of disciplines. Continuing to hone the ability to effectively articulate both the goals of innovative assessment tools and approaches and the results of assessments will facilitate needed cross-cutting R&D.

2.2.4. Future R&D

A three-pronged approach to focusing and sustaining R&D targeted at developing innovative new approaches and applying them to high-priority technologies, systems, and architectures will address both the needs and challenges identified. Effective integration between multiple programs and organizations (e.g., NE, NNSA, NRC, and the Department of State) will be an important part of executing needed R&D. Drawing on a rigorous and expanded scientific base will be essential for success. This will include research by institutions of higher education, national laboratories, and other organizations.

- Develop a proliferation and terrorism risk assessment RD&D roadmap.
  - Include a cross-section of potential areas for new and innovative work.
  - Utilize independent expert review and advice.
  - Incorporate results of fundamental R&D and advanced applications.
- Perform science-based investigator-driven fundamental R&D.
  - Develop innovative methods.
  - Utilize cross-disciplinary teams (including, e.g., political scientists, social scientists).
  - Focus initially on key topics and approaches that will significantly advance capabilities, for example:
    - Adversary decision models for proliferation and terrorism
    - Mathematical methods for performing risk analysis (e.g., game theory)
    - Treatment of uncertainty
    - Best practices for expert elicitation
    - Risk communication
  - Enable assessment of multiple pathways in a systematic manner with consideration of:
    - Pathway attributes affecting adversary choices (e.g., attractiveness, cost)
    - Probability of detection

        - Delay after detection
        - Latency
        - Consequences (including effects beyond nuclear detonations and radiological dispersion)
    - Improve transparency and repeatability of analyses.
- Apply risk assessment tools to high-priority problems and issues.
    - Establish standardized "benchmark problems" for consistent comparisons of different methods.
    - Perform prototypic evaluation studies of proliferation and terrorism risks, focusing on a range of nuclear energy systems of interest to DOE.
    - Demonstrate the value and uses of risk assessments; understand their practical strengths, limitations, and the insights that can inform risk management decisions concerning advanced nuclear energy systems.
    - Highlight gaps in risk assessment methodology to inform the longer-term RD&D agenda.
    - Provide guidance on metrics that can be used by systems analysts when considering (in an approximate way) key aspects of proliferation and terrorism in context or other parameters such as economics, safety, and waste management.

Ultimately these R&D efforts will lead to both evolutionary and revolutionary tools and methodologies for assessing and comparing the proliferation and terrorism risks of alternative nuclear energy technologies, nuclear fuel cycle systems, and global nuclear energy architecture options. These new and improved tools for proliferation and terrorism risk assessment will strengthen our ability to perform grounded comparison of the benefits and drawbacks provided by different options. These tools and methodologies will also provide a yardstick for measuring progress in addressing proliferation and terrorism issues and/or barriers, and ultimately in producing a set of innovative and viable options for the future nuclear energy enterprise.

2.3. Reactor Materials

This session covered the component of the program that will facilitate new classes of alloys and materials through openly competed alloy development efforts and promotion of modern materials science tools and techniques. This will enable transformational reactor performance under a wide range of operating conditions with the general goal of improved strength and radiation resistance. Innovative steels designed with modern materials science techniques may increase maximum operating temperature as much as 200°C and high dose limits by 100 percent over the leading materials systems.

The session had four main goals:

- Describe the proposed scope of activities related to innovative reactor materials and their relevance to DOE-sponsored programs.
- Delineate between proposed technical activities and program elements related to innovative reactor materials in other DOE-sponsored R&D programs.
- Discuss input on gaps between needed and existing capabilities.
- Identify areas for coordinated research between this proposed R&D program and other R&D efforts.

2.3.1. Overview/Current State of the Art

Materials degradation and performance is a common concern in existing reactors, and understanding the long-term behavior of materials in the reactor core, vessel, and other subsystems is critical for safe and reliable reactor operation. Little data exists on the dynamics of material degradation and failure modes in harsh environments. For the existing nuclear reactor fleet, areas in which materials issues must be resolved include:

- Reactor pressure vessels (RPVs) and primary piping
- Core internals
- Secondary systems
- Weldments
- Concrete
- Cabling
- Buried piping

Some current issues in materials technology, according to industry experts, include:

- No integrated models for long-term behaviors
- Limited range of accuracy
- Limited range of extrapolation for material systems (for example, there is a possibility of late-blooming phases in RPV steels)

- Understanding of processing and chemistry variable behavior
- The possibility that current practices use excessive conservatism in some cases
- An inability to differentiate between certain materials
- Limited acceptance of alternative technologies, which require significant validation and qualification efforts

Some LWR materials issues are being examined in other DOE programs. For example, current reactors with Zircaloy fuel cladding, stainless steel reactor internals and ferritic steel pressure vessels have certain materials challenges. Fuel behavior under more severe duty leads to acceleration (breakaway) of processes such as irradiation growth and uniform corrosion. Also, hydriding can cause embrittlement and degradation of reactivity-initiated accident (RIA) resistance. Pressure vessel embrittlement is an issue for plant life extension. Stress corrosion cracking (SCC) of reactor internals is also a problem.

One current mission in the DOE materials research programs with regard to innovative reactor core materials is to develop and demonstrate fabrication processes and in-pile performance of advanced fuels and targets (including the cladding) to support the different fuel cycle options defined in the NE roadmap.

Advanced metallic alloys and even ceramics are being researched for possible use in the fuel cycle. In certain ceramics such as SiC, radiation-induced–ultra-fine–high-energy defect nanostructures make the ceramic material extremely resistant to radiation. Recently SiC was proven to be subject to progressive radiation damage beyond 3 dpa up to ~40 dpa. In certain metallic alloys, such as ODS steels, the fine oxide dispersions have shown resistance to radiation doses greater than 100 dpa.

Despite differences in operating requirements and conditions, there are common needs across different goals and technologies and programs. Alloy development, for example, is a common activity among key programs. Joining is universally required, although in different forms. Compatibility, radiation damage and predictive capabilities are also applicable to various research efforts. Finally, ASME code qualification and subsequent NRC approval are important for all reactor technologies.

Cutting-edge research in modern materials science requires new tools: new microscopy techniques, new characterization tools, new computational ideas and applications.

Today, many nuclear engineering problems require interface modeling, fluid instability, phase transition, fluid–structure interaction, etc., coupled with various systems for simulations of nuclear engineering problems. For example, the pair distribution function (PDF) technique provides information about the atomic arrangements over short, intermediate and long distances. PDF was originally developed for liquids and amorphous materials. However, its ability to characterize materials on various length scales makes it ideal for studying nanomaterials. Thanks to advances in computational modeling and high-energy synchrotron sources, PDF can be used successfully on more materials. High-resolution energy dispersive x-ray diffraction (EDXRD) and phase mapping can be used to measure the internal stress/strain field or map phases and chemical distributions within samples. EDXRD is useful for engineering applications, particularly in examining phase and/or chemical composition over time, such as assessing hardening cement or mapping residual stress from shot peening or fatigue cracks.

As noted above, addressing radiation damage is one common need for most materials and reactor systems. However, neutron irradiation sources, while prototypic, are expensive and sparse. These facilities are being used for long-term irradiations, but it takes many years to perform the irradiations, and the post-irradiation examinations (PIEs) require hot cells extending the time for analysis for another one to two years. Currently, simulations of specific high-dose irradiation conditions are being investigated using ion irradiation. One novel technique uses *in situ* transmission electron microscopy (TEM) analysis of materials under ion irradiation to study the fundamentals of defect formation under irradiation.

The combination of multiple ion beams has been used historically to perform controlled fundamental studies of radiation effects on materials at facilities worldwide. Several facilities in Europe and Japan combine multiple ion beams with TEM. However, a facility at a synchrotron or a high-energy proton accelerator combined with *in situ* characterization techniques would be unique in the world. Emulating specific radiation effects in a reactor environment would require an ion source in the hundreds-of-kilovolts energy range (ion implanters) for implanting helium and hydrogen and an ion source in the megavolt range accelerating ions of the same material as the sample (self ions). Some current issues in multiple ion beams research follow:

- Computational modelers are frequently required to try to decipher complicated and path-dependent material processes from very few data points for macroscopic materials properties at the end of long irradiation exposures.

- Experiments at far-from-nominal conditions are generally not performed, but could be beneficial to computational modelers and provide valuable data so that materials performance estimates are not solely based upon extrapolation during reactivity and temperature transients.

The new tools and research targets provide opportunities for increased collaboration, such as enhanced collaboration within DOE–NE programs, new opportunities with other DOE offices, and the creation of new mechanisms for collaboration with non-government entities.

2.3.2. Future State and Knowledge Gaps

As the nuclear power industry moves forward with building new LWR plants, extending the lifetimes, or developing new advanced systems and investigating options to close the fuel cycle, several key needs and directions will be required. Each is described below.

There are several materials technology needs associated with the upcoming increase in demand for nuclear power, as well as for continuing support of current nuclear power-generating capability. Different reactor technologies are being considered to fulfill upcoming power demands in the near term, intermediate term and long term. Table 1 lists the material implications of these technologies.

Materials Performance Issues. The nuclear community requires a well-validated extensive database of materials properties to reliably predict and overcome materials degradation issues. Existing materials will need to be put into new regimes, and testing must reach the same level at which existing materials have been validated. Desirable properties for fast reactor clad and duct (Figure 4) include:

- High thermal conductivity (allows lower fuel temperature)
- Compatibility with fuel (especially for metal fuels)
- Compatibility with coolant (typically sodium)
- Sufficient strength and ductility
- Thermal creep resistance (allows higher coolant ΔT through the reactor, which leads to greater thermal efficiency for power)

- Physical and mechanical immunity to radiation
- Affordability

Table 1. Materials Implications of Near-Term, Intermediate-Term, and Long-Term Technologies.

	Near Term	Intermediate Term	Long Term
Plant Options	Relicense existing Gen III plants (40, 60 years)	Relicense existing Gen III plants (60, 80 years) Build new Gen III+ plants	Relicense existing Gen III+ plants (80 years) Build new Gen IV plants
Reactor Techno-logies	Existing BWR and PWR plants	Advanced BWR and PWR plants	Improved BWR and PWR plants, supercritical water reactors, gas-cooled reactors, lead- and sodium-cooled reactors
Materials Technology Implications	Understanding of existing plant materials Quantification of time- dependent materials properties	Extended unders-tanding of traditional materials validation of "improved materials" used to repair Gen III reactors	Understanding of materials properties in extended conditions (temperature, environment, etc.)

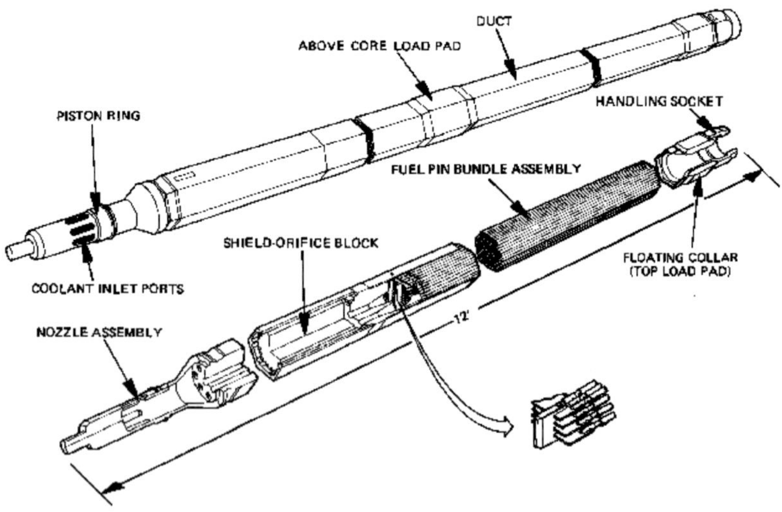

Figure 4. Fast reactor clad and duct.

Having a knowledge base of the suitability of materials for specific nuclear plant applications is the future state of the art in materials research. The database should benchmark computational tools with real measurements.

In addition, there is a fundamental lack of knowledge in the processing–microstructure–property relationship in advanced materials (steel, concrete, and non-conventional reactor materials). Validation and trust in new testing techniques and research-level materials is also important. Also, obtaining real-time, continuous monitoring data of material corrosion and degradation in extreme environments is necessary to achieve many of the stated reactor materials goals. Real- time studies have not been applied to materials in extreme environments.

Extended Service. New materials challenges caused by extension of service life to 40 years, 60 years or beyond include increased exposure to extreme temperatures, stresses, coolants and neutrons. Choosing the right materials can impact key flexibility, safety and economic requirements for advanced fast reactor development.

Plant life extension will also require certification that material degradation has not affected the plant's ability to perform safely, especially with regard to pressure vessel embrittlement. Higher fuel burnup, which decreases waste volume and allows longer fuel cycles, would use more highly enriched fuel (five percent) and longer residence times. This leads to increased radiation damage and corrosion. High burnup also affects accident limits for RIAs and loss-of-coolant accidents. Power uprates increase outlet coolant temperature to bump up power, leading to higher corrosion rates and flux gradients. Degradation of fuel during dry storage is another issue.

Some key materials technology needs for LWR plant sustainability are in the areas of RPV steels, irradiated stainless steels for internals and nickel-base alloys. With regard to irradiated stainless steels for LWR internals, plant relicensing will call for quantitative degradation models of plants for specific configurations and service histories. Relicensing has demonstrated the capability for advanced finite element analysis (FEA) and materials modeling to provide useful, quantitative predictions of materials degradation. This research area must also investigate measuring the effects of aging and dose on certain components and understanding material variants.

For LWR RPVs, drivers and opportunities in materials research include refinement of the embrittlement trend curve analysis, which the extension of reactor life to 60 years necessitates. Life extension also calls for high fluence and understanding of flux rate effects. High fluence data entails a number of additional requirements. Also, the use of an alternative pressurized thermal shock rule and trend correlations calls for assessment of applications to specific plants with specific data. These drivers and opportunities are applicable to new plants as well as existing ones.

Advanced Materials Development. Regarding fuels, enabling multifold increases in burnup will require development and testing of advanced alloys suitable for clad and duct and other high-dose core components to >400 dpa over the clad and duct operating conditions. These alloys must be radiation-tolerant, corrosion-resistant, weldable and processed into tube form— and cladding must be resistant to chemical interaction with fuel.

Using FM steels for cladding creates issues with increased radiation damage that can cause microstructure evolution in cladding property degradation; this is especially true for sodium fast reactors. The higher temperature also requires corrosion-resistant material, especially with supercritical water-cooled reactors and lead-cooled fast reactors.

Development of either radiation-tolerant materials or radiation-controlled materials will be necessary. Examples include extremely radiation-tolerant ODS steels or innovative new materials such as alloys that precipitate radiation-tolerant precipitates under irradiation or amorphous metallic glasses. A desired future state could entail development of radiation tolerance through grain boundary/interface engineering (e.g., nanograined Cr/Nb multilayers). This would lead to radiation-controlled material (i.e., a multilayer that has a weak layer that breaks off after given exposure, possibly with built-in sensors of material properties).

Composites are likely to greatly advance materials for nuclear reactors. They break the link between properties usually used in conventional homogeneous materials, such as ductility and strength. The challenge in the nuclear industry will be acceptance because the reactor must be designed differently to enable use of materials with very low fracture toughness and difficult joining challenges. Designers may not be open to considering composites.

Understanding radiation-induced radiation insensitivity in ceramics is important. In certain ceramics such as SiC, defect configuration and the stabilizing mechanisms have yet to be understood. It is necessary to achieve understanding of how SiC accommodates one-percent Frenkel defects that are stable and finely distributed. This may lead to the finding of other materials or materials systems that are inherently radiation-immune. Tiny defect clusters, antisite defects, defect-impurity clusters, etc., are not quantitatively characterized presently.

Modern Materials Science Tools. Combining x-ray total diffraction and PDF analysis of data and EDXRD, it will be possible to evaluate radiation damage and measure material properties on the same sample. This will be a powerful and unprecedented opportunity to investigate the effect of radiation damage on material properties. Relating microstructural changes to material properties is essential to the materials design process. By combining these experimental capabilities, it may be possible to directly relate the environmental conditions (dpa and annealing temperature) to microstructural features (dislocation structures, point defects, segregation of solutes) to mechanical properties (elastic modulus and yield strength).

With regards to multiple ion beam research, a breakthrough that needs to be pursued is nondestructive 3D diffraction probes with micron-range resolution, which can provide information on positions, orientation, boundaries, stress and strain, size and phase change of each individual grain during irradiation. Another breakthrough is the characterization on mesoscopic length scales to provide a linkage between micro- and macro-structures (coupling small-angle and wide- angle x-ray scattering [SAXS and WAXS]). The new generation x-ray tools open the way to time-resolved measurements: metastable and transient states can be looked at, and observations can be carried out as the material evolves under the irradiation/implantation beam. For example, fast CMOS area detectors can collect 10–1,000 frames per second while the material is being irradiated.

2.3.3. NEET Materials Cross-Cut Vision

The advanced reactor materials cross-cut effort will enable the development of innovative and revolutionary materials and provide broad-based modern materials science support to ongoing materials research within all of the NE imperatives. This will be accomplished through innovative and competitively awarded materials development, promoting the use of modern materials science via up-front investments in new capabilities and establishing new research partnerships with other agencies.

Under this effort, innovative materials development is a key element. This competitively awarded research will support all nuclear concepts and, if successful, enable revolutionary new reactor designs and performance limits. In several programs (Fuel Cycle R&D, NGNP, LWRS, Gen IV), elements of similar research already exist, but these are mostly incremental in nature and are specific to one or a limited number of applications. Research into specific degradation modes or materials needs unique to a particular reactor design should be performed in the supporting program as specific requirements and unique testing needs; potential for overlap makes inclusion in the materials cross-cut less advantageous. The NEET cross-cut research offers much higher risk and reward.

To support these objectives and research within all the NE imperatives, three major research areas are envisioned as part of the materials crosscut. Specific FY 2012 goals include:

- Initiating competitively awarded research tasks on innovative and enabling materials.
- Providing modern materials science tools to this research and all imperatives.
    - Competing and acquiring at least two new tools for national research.
    - Enabling and enhancing the nation's capability for ion beam irradiations.
- Improving cooperation and collaborative research for nuclear materials.

Each of these primary areas is described in more detail below.

Innovative and Enabling Materials. Using modern materials science techniques and research partnerships, this task will design, develop, and validate new structural materials or fabrication techniques for use in advanced reactors supporting all NE imperatives. New high-risk–highreward concepts well beyond those currently considered by most industrial interests will be explored to provide alloys with improved performance over traditional materials. Improved performance may include a five- to ten-fold increase in strength or increased maximum operating temperature by over 200ºC with a service period of at least 80 years. Concepts such as optimized steels with engineered microstructures may provide revolutionary gains. Other more radical concepts such as multi-layered metals or metal–ceramic composites have great potential but have never been evaluated for nuclear service. New concepts will be explored via a competitive proposal and seed-funding process. After a three-year period, the most promising concepts will be selected for further development, and a new batch of seed projects will be chosen. Specific criteria for award selection will be developed in FY 2010 and based partly upon an upcoming Gordon conference on extreme materials.

Modern Materials Science. Upfront investments in new tools will enhance the application of modern materials science techniques to research in the five NE imperatives and major programs. Key components to modern materials science include computational techniques, experience, and modern tools and research techniques. Investments in new tools such as atom-probe tomography, high-resolution mapping tools, and other techniques such as ion-beam radiation or new tools for radiation damage studies will promote greater gains in all areas of research.

Collaborative Materials Research. Fostering new partnerships with other agencies may yield new research partnerships, enhanced collaboration, or possibly shared investment in research facilities. Possible collaborators include the Office of Science, the Office of Fossil Energy, the Office of Fusion Energy Sciences, the ITER project, EERE and NASA. Increased visibility of NE programs will spur interest in new partnerships.

2.3.4. Future R&D

- A number of industrial solutions can support sustainability in the power-generating reactor fleet. They are:
  - Application-specific empirical models (PWR, BWR, Fast Reactor)
  - Specific alloy performance data
  - Bounding data for material classes
  - Sparse database
  - Materials repair and mitigation processes (proof-of-performance data)
  - In-place NDE methods and performance monitoring
- Some basic science support is needed for these efforts, particularly in the following areas:
  - Integrated quantitative models of materials behavior and degradation
  - Metallurgical effects in alloy behavior (chemistry, microstructure effects)
  - Property variation with respect to material variants
  - Extension of materials property database

    - New materials repair, coating and surfacing methodologies
    - Advanced sensing and monitoring technologies – application to "materials condition"
- Some drivers and opportunities have been identified in materials science that will be needed for future reactors. Understanding is needed in these areas:
    - Materials properties in new environments and temperatures, such as different cooling systems (gas, liquid metal) and temperature increases
    - Thermal irradiation interactions (mechanisms)
    - Rates of interactions
    - Quantitative assessments of materials property retention (reverse of degradation)
- To ensure good materials knowledge in time for the development of future reactor designs, the research community needs to:
    - Target materials performance values
    - Identify systems options
    - Generate data
    - Build models
    - Produce reliable predictions
- With regard to innovative core materials, the nuclear community must:
    - Develop fuels and targets that increase efficiency of nuclear energy production
    - Maximize utilization of natural resources
    - Minimize generation of high-level waste
    - Minimize risk of nuclear proliferation
- The great challenge will be to obtain:
    - Multi-fold increase in fuel burnup over the currently known technologies
    - Multi-fold decrease in fabrication losses with highly efficient predictable and repeatable processes
- In certain advanced materials such as SiC, ODS steels, amorphous metals, and other advanced alloys, future R&D needs include:
    - Advanced characterization of defects in irradiated materials, e.g. positron, LEAP, semi-conducting proper characterization
    - Advanced *in situ* characterization in TEM, ion beams
    - Atomistic computational simulation and analysis

- Development of fully integrated nuclear power plant control systems is also important to reactor materials development. Researchers need to:
  - Develop a rapid method to characterize the microstructure and mechanical properties of various new alloy compositions at various dpa levels
  - Perform rapid testing of a large number of materials up to high dpa levels
  - Develop a microscope with nanometer resolution and nanosecond-to-microsecond time resolution for *in situ* irradiation
- Drivers and opportunities in the area of nickel-base alloys for LWRs include:
  - Forged alloy 690 behavior (SCC initiation data and SCC crack propagation rates)
  - Alloy 690 weldments (SCC properties of weldments)
  - Chemical mitigation effects (crack initiation and growth, hydrogen, zinc, noble metals)
- Researchers must characterize radiation damage through:
  - Total x-ray diffraction with subsequent PDF analysis
  - High-resolution EDXRD and phase mapping
  - Observation of annealing of radiation damage *in situ* at a beamline
  - Study of the interactions of defects with interfaces internal to the microstructure
- Researchers should carry out real-time studies of material transformation under conditions relevant to nuclear plant operations using methods such as:
  - Active radiometry (corrosion)
  - Interferometry (swelling)
  - Plasma sampling (surface chemistry)
- PDFs and structure determination need to be performed for materials of interest for advanced reactors and nanomaterials. Knowledge gained from these studies needs to be incorporated more quickly into the materials design cycle. Fabrication, irradiation testing, and characterization should be performed iteratively to optimize material microstructure to withstand high-temperature–high-stress–high-radiation environments.

- With regards to research on multiple ion beams, future R&D should include:
  - Measurements at existing beamlines on irradiated samples to define dose and implantation depth versus the beamline parameters and techniques used (energy, beam size, flux, resolution)
  - Specification of the ion beams
  - Simulations of beamline performance
- There is currently no fast neutron source for testing in the United States. Researchers must go to China or Japan for that capability. DOE may have to rely on shielded radiation and special capsules in the Advanced Test Reactor, as well as other tests, which will buy time to get access to international facilities.

2.4. Advanced Sensors and Instrumentation

This workshop session covered sensors, instrumentation, and related technologies that crosscut nuclear energy programs. Topics discussed included advanced sensors, instrumentation and algorithms for advanced diagnostics and prognostics to improve physical measurement accuracy, predict future state, and both quantify and reduce uncertainties; digital monitoring and control technology; fiber optic and wireless digital instruments; and integrated control systems and decision aids to improve performance and reliability.

2.4.1. Overview/Current State of the Art

The sensors and instrumentation in existing plants were built to ensure safety in operation; many elements of these systems are not optimized for human–system interaction. Nor were they designed to facilitate optimizing operations, performing maintenance, modernizing systems and enhancing functionality. Current LWRs have individual control centers that are functionally dedicated, physically isolated, largely analog, and spatially relatively large. Operational activities are human-intensive with little automation outside safety and protection systems. Over time, staff activities to maintain the status quo of systems have increased for a number of reasons, including:

- Increased effort and costs to maintain I&C systems that are aging and becoming obsolete

- Reduced staff and knowledge drain (e.g., staff aging and retirement, loss of tacit knowledge, new plants thinning talent)
- New requirements and commitments such as for addressing cyber security concerns

Overall there are few computer-based decision and support aids to help staff perform activities more efficiently to reduce burdensome workload and stress levels, and reduce likelihood of human errors. More automation would allow for complementary capabilities to be implemented at the control system level to enhance monitoring of process variables and implement control actions; increase system status awareness, reliability and availability; and reduce staffing for functions that can be automated and completed more efficiently using computer-based capabilities.

At current plants, LWRs primarily use analog technology for control and safety instrumentation in stove-piped system architectures employing conventional control theory (i.e., proportional–integral–derivative [PID] control). The use of digital technology for safety-related functions has been limited with deployment of most digital modernization activities focused on high pay-back controls (e.g., turbine, feedwater). Current instrumentation capabilities do not monitor or examine risk-critical components and functions to automatically capture and process data relating to those functions. Condition-based maintenance still predominately uses periodic manual measurements. Regarding fleet-wide monitoring and diagnostics, there has been limited adoption by the nuclear industry. Fleet-wide monitoring has found moderate penetration into the fossil industry with proven success.

Regarding resilience, reactor protection systems depend on very limited setpoint modification, preventing reactors from supporting grid restart after a blackout. Risk meters support nuclear power plant configuration management using an underlying PRA model. Risk meter technology helps ensure that in planning outages the facility remains within an envelope of acceptable risk.

For condition-based monitoring (CBM), both active and passive component aging requires awareness through measurement, and the associated sensor requirements are different than for operational measurements. Passive components (e.g., pressure vessel, class one piping, etc.) and their degradation mechanisms may be expected to increasingly dominate risk as a plant ages. Active components (e.g., pumps, valves, motors, etc.) are likely to be largely managed by CBM and replaced in a timely fashion (i.e., just in time) at the onset of critical degradation. Recent advances have begun to focus on the means of incorporating specific materials degradation mechanisms into PRA models, which provide the engine for risk meter technologies. There is, as yet, no strong methodological basis for the risk analysis of passives, although work is in progress in this area.

Current sensor technology consists mostly of devices that generate analog signals, when accuracy and reliability might be improved through introduction of available digital technologies. Known issues exist relating to drift and measurement uncertainty for some families of sensors. This situation constrains plant staff activities through increased inspection and outage testing intervals, required regular manual calibration and *in situ* testing during power operations, all of which are costly, labor-intensive and time-consuming. To minimize cable runs, which are costly to install, maintain and replace, hardwired power and data cables employ multiplexing and localized electronics to the extent practical. However, backfitting additional wired sensors is very costly and inhibits introduction of new measurement capabilities to support enhanced plant status awareness and condition monitoring.

Given the number of sensors and monitors at any given nuclear power plant, a wealth of information is constantly being generated and transferred through cabling and limited networking. Since the last new nuclear plant was built in the United States, communication technologies have greatly advanced. However, NRC has yet to approve current state-of-the-art technologies, such as digital communication. At existing plants, automation through the adoption of digital technologies tends to be found in islands. Small parts of the plant often communicate very well within their respective areas but not very well with the entire plant. Additionally, data transmission is limited by unique environments of advanced reactors, such as high temperatures and radiation. Finally, sensors and instrumentation systems and associated systems are often "paste-ons" in the design process.

Current practice in the U.S. nuclear industry is considerably behind other process industries in the use of digital technology. A significant factor contributing to this condition is the uncertainty and perceived financial risk arising from the current subjective regulatory framework for I&C systems at nuclear power plants. The primary issue that complicates the treatment of digital I&C systems within licensing assessments is uncertainty about how to address software. The complexity of software-based systems drives uncertainty in understanding failure mechanisms, estimating reliability, and ensuring quality. Consequently, subjective regulatory reviews introduce significant conservatism to provide reasonable assurance of safety, thus imposing a cost and predictability burden on the licensing process that inhibits the use of modern technology. Two key manifestations of regulatory uncertainty involve software quality assurance and diversity and defense-in-depth. While the regulatory situation for software quality is stable, it depends on a costly process-oriented approach to establishing quality that still leads to subjective rather than objective evidence. Thus the quality of software-based systems cannot be quantified or readily assessed. The present inability to establish a clear basis for the degree and nature of diversity that is necessary to resolve concerns about CCF is the highest-priority I&C issue facing the nuclear industry today. Current approaches to satisfy regulatory concerns are complex and of uncertain effect. The combined impact of these technical issues has contributed to delay or cancellation of modernization projects while introducing a critical path concern for the licensing of new plants.

2.4.2. Future State

As with the current deployed instrumentation technology, safety is by far the most important factor when designing a new plant. However, the next generation of reactors will probably also consider the impact of design features on effectiveness of human performance. Advanced concepts for nuclear reactor designs will include more highly automated control and information systems and therefore have the potential for a reduced operational staff. New plants will have new I&C architectures, with increased functionality, for distributed control, remote operation, and remote expert support. These new operation technologies will integrate automation and human performance for safe and efficient operation.

The future of operations will incorporate changes to concepts of operation with highly integrated control and display capabilities, integration with centralized expertise and fleet-wide and/or industry-wide asset support functions. The use of computerized aids, intelligent agents, automation and remote experts will help to reduce operational staff levels. There will be shared control through function allocation and adaptive automation. Distributed control systems, possibly employing intelligent software agents, will communicate with smart instruments and controllers. Real-time data mining will be available with faster-than-real-time monitoring (model-based) and much-faster-than-real-time scenario simulation. These advances will allow operators to investigate multiple scenarios before acting on a decision. Advanced human–system interface (HSI) technologies will include interactive interfaces. Computerized decision aids and displays based on cognitive research will be used to support better and faster operator decisions. Operators will also have improved situation awareness, knowledge capture and transfer approaches.

Future plants will have resilient and adaptive digital controls. These new controls will be capable of functioning in high-noise (uncertain) environments. They will have the ability to account for system conditions and configurations to function with fewer mode dependencies, and they will be able to adapt their control algorithms to avoid over-or under-control. Figure 5 shows a performance of resilient communication graph. Truly resilient power stations will withstand grid blackouts. Plants will also have centralized, nearly autonomous multi-unit controls with bump-less cogeneration capabilities and with command and control validation. Improved monitoring will allow for optimized operations, maintenance, and asset management decisions by providing integrated detection, diagnostics, and prognostics. There will be plant- and fleet-wide integration of asset information. A dynamic plant-wide control system could also communicate with local emergency agencies, other power plants, and electric grid operators.

Future risk meters or systems that provide a system condition index could analyze real-time data from a network of active and passive element monitors to provide a dynamic, predictive condition/risk profile for key systems and sub-systems and units of the plant. This methodological base would also provide risk-informed insight into design and placement of advanced sensors to achieve the greatest uncertainty-reduction efficacy in risk prediction.

Sensors will be developed to withstand higher temperature regimes and harsh environments for new operating conditions and applications. New sensors will have the ability to directly measure primary process parameters that would otherwise be inferred or measured from a distance, thereby resolving the technical challenges posed by corresponding loss in precision and increase in uncertainty that limits the value of indirect measurements for control and condition prediction. Sensors could be integrated into key active and passive material elements (e.g., for cladding damage measurement). RFID tags applied to nuclear material could be used for security. Future sensors will be able to give enhanced system condition awareness. For example, new sensors can provide accurate measurement of circulating flow, activity levels in coolants, and even real-time measurement of the purity of working fluids, which can enable proactive management of coolant chemistry and, in turn, help minimize corrosion of pressure boundary materials and components that contact the fluids. Sensor quality will also be improved, minimizing measurement drift and supporting longer intervals between maintenance and service outages. As sensors improve, so must their associated electronic components. Electronics will also need to be radiation-tolerant to enable installation in close proximity to the process (e.g., in-vessel) and support localized processing of low-amplitude, high-noise measurements.

To support these innovations, future plants will require new methods of communication. They must be able to support greater data generation, higher sampling frequencies and transmission demands that are expected to accompany advancements in digital sensor, measurement, and control technologies. These new methods must be qualified for the unique operational environments represented by different advanced nuclear energy programs. Finally, any new form of communication must meet standards and regulatory requirements for data integrity, reliability and security.

Future plants will likely take advantage of wireless communications, which currently have reliability and security concerns. An operator relies on uninterrupted signals to make informed decisions, so resilient communications will be used. They will be secure and open and have early perception and response times, signal integrity, and state awareness.

Barriers and unnecessary regulatory burdens will be resolved for optimal, cost-effective use of advanced technology at future nuclear power plants. In effect, the regulatory framework for licensing digital I&C systems will be based on objective rather than subjective acceptance criteria. The reduction of uncertainty will result from research that investigates the nature of software-based systems and provides the basis for characterizing quality and dependability (i.e., safety, reliability, availability, diversity and robustness to faults) in terms of quantifiable measures through well-established metrics.

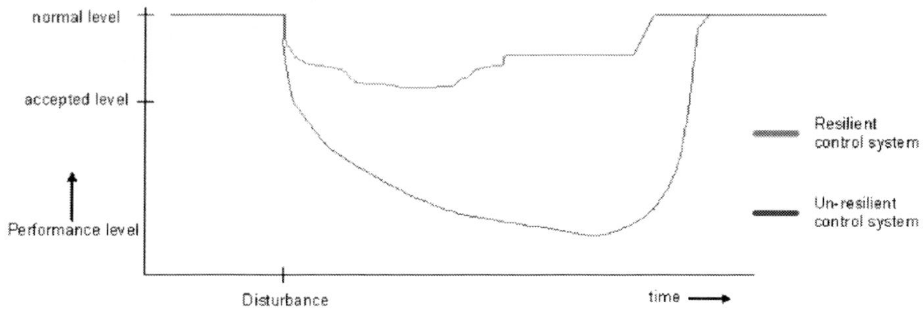

Figure 5. Performance of resilient communications.

2.4.3. Knowledge Gaps

There are gaps between our knowledge of the current state of operations and that expected for the future state of operations. New concepts of operations will be developed which will require testing and validation. Highly integrated controls and displays supporting plant and human cognition and response are needed to allow for effective teamwork between human and automated systems. We also need to know that operators will be able to effectively intervene if automation or computer-based aids fail. To help humans integrate with automation, physical and virtual reality control room test beds must be built to help develop and test proposed solutions and design alternatives cost effectively. Cognitive research is needed for decision making and information displays, along with much-faster-than-real-time scenario simulation. Effective and secure collaboration with remote experts will allow for reduced staffing without negative impacts on safety and performance. Verification and validation and configuration management of new technologies must be developed. Effective procedures and training approaches for new technologies and concepts of operations will need to be created.

Proposed control technologies—autonomous, reconfigurable, resilient, and multi-module—must be developed and demonstrated. There is a need for well-defined proof-of-concept benchmarks and studies that could be performed at dedicated testing facilities. Progress requires a science and technology knowledge base for on-line measurements that enables implementation of diagnostics and prognostics, to either supplement or potentially replace some current in-service inspections. As system capabilities are enhanced, sensors must have a wider range of measurement capabilities, as well as better data acquisition and transmission, with increased bandwidth requirements, to facilitate the use of data in diagnostics and prognostics. When condition assessments or indices and remaining useful life metrics are calculated, the model-based prognostics include uncertainty in resulting data. It is necessary to first understand and then bound what are inherently ill-posed inverse problems with sparse data that, in many cases, are poorly constrained. Determining confidence measures for the prognostic estimates (ideally bounding the worst-case uncertainty) is a significant mathematical challenge.

The nuclear sensor industry needs to develop and qualify new sensors, addressing environmental robustness through the materials employed and demonstrating innovative measurement methods to detect and monitor physical condition and behavior of plant systems. Achieving the future state will require innovative measurement and data integration/predictive technologies, robust materials to withstand high temperatures and radiation, improved sensor functionality, new standards, and test bed facilities. The industry lacks an understanding of what really needs to be measured to ensure future plant safety, as well as the rate at which change occurs (i.e., the remaining useful life). Safety and capacity factors could be maintained and potentially improved with new sensors, which should also have self-diagnostic capabilities and promote data integration to enhance condition and future condition assessments.

Each new reactor design will have its own specific measurements needs. Sensors and related instrumentation will have to be tailored for each type of new system. Process measurements need to be improved in terms of uncertainty and quantification. Self-calibrating and drift-free sensors are needed for extended as-is operations. Suitability for deployment within an in-vessel environment is necessary to support continuous long-term operation with minimized penetrations. Research is needed on material development and testing of sensors and to support qualification for harsh environments. In addition, innovative measurement technologies are lacking for direct measurement of process variables or health/condition indicators and for assessment of future condition.

To support the new systems, the nuclear industry must have communication technologies with proven:

- Physical and cyber security
- Safeguards to support nonproliferation
- Process stability and efficiency
- Human and automation considerations
- Interleaving of human and automation responses
- Automation on the fly, semi-autonomous operations
- Measurement and adaptation considerations

How the information is presented to operators must also be addressed. This factor includes the collation of diverse indicators, prioritization of information, and tailoring of consumer information. Future communication technologies will need a hierarchical, multi-agent approach, which can also provide independent oversight.

As with any new technology or product, these systems will have to meet the proper requirements, guidance and regulations. In some cases, specifications and approved regulatory guidance must be updated so that the nuclear industry can use these new products. Specifically, developing and demonstrating comprehensive quantitative measures of software quality and digital system reliability are needed to address regulatory uncertainty. In addition, optimal resolution of current high-priority licensing challenges requires determination of effective, technology-based strategies for implementing diversity and defense-in-depth within the I&C architecture of a nuclear power plant to replace expedient, overly conservative strategies that are currently emerging. Research to resolve these technical issues holds promise of potential transformative breakthroughs in the treatment of digital technology that can improve cost, reduce complexity and regulatory uncertainty, and eventually remove impediments to the modernization of I&C technology usage in the nuclear power industry.

2.4.4. Future R&D

To improve operations, researchers need to develop/test/demonstrate:

- Automation and intelligent aiding to support a target staffing regime, including utilization of off-site expertise and support to a reduced staff control room
- Advanced automation and shared control technologies for different advanced nuclear energy programs
- Advanced HSI information and communications technologies to support monitoring and control, including impact on human performance
- Performance measures for control systems and operational staff to assess the impact of staffing changes and automation on human performance
- A technology-neutral approach for licensing, including development of an objective technical basis for resolving key issues such as:
    - Diversity (nature, impact and comparative value for I&C systems)
    - System dependability (including metrics for software quality, models for digital system reliability, and testing approaches to validate fault coverage)
- Expanded HSI and controls to include new concepts of operations:
    - Display of systems associated with equipment being maintained (electrical, fluid, control)
    - Adaptive control
    - Prohibited actions (trips, technical specification violation, personnel safety)
    - Integrated on-line monitoring and diagnosis
    - Resilient control
    - Technical specification limit timer

A designated facility is needed for testing and demonstrating advanced monitoring and control techniques, as well as prognostic testing. To improve control techniques, researchers need to develop/test/demonstrate:

- Autonomous and reconfigurable controls
- Cyber security in remote control/operation
- Resilient control strategy for decision-making and control under unknown internal or external causes
- Functional architectures to enable integration of controls, diagnostics, and decision capabilities in support of autonomous operation
- An understanding of the control of multi-modular nuclear stations
- A risk-informed instrumentation process
- Dynamic setpoints for reactor protection systems to improve resilience
- Robotics

Monitoring, diagnostics- and prognostics-related R&D needs include:

- Identifying critical equipment and differences between active and passive system elements
- Quantifying prognostics uncertainty
- Defining data sets to facilitate trials for testing prognostic algorithms
- Developing integrated life-cycle diagnostics and prognostics algorithms
- Integrating diagnostic and prognostic information into dynamic PRAs
- Identifying and modeling dominant failure modes for critical equipment
- Designing and evaluating sensors necessary for degradation monitoring
- Developing sensor-failure-tolerant diagnostics and prognostics algorithms
- Finding novel ways to use sensors
    - Noise analysis
- Integrating data from local inspections into a global condition assessment
    - Fundamental differences between operating an aging plant using periodic in-service inspections and deploying online monitoring

- Using multisensor data fusion for local and global condition index estimation
- Determining confidence measures for prognostic estimates (ideally bounding the worst- case uncertainty)
- Coupling controls with models including FEA to identify potential high-fatigue evolutions and other parameters to be monitored
- Developing acoustic monitors that listen for cracks
- Developing advanced image and vision systems to:
    - Enforce the two-person rule
    - Identify intent by tracking movement within a facility
    - Remotely monitor processes and in-vessel components
- Investigating sensor placement to facilitate effective implementation of control and diagnostics

Other R&D specific to sensors should examine:

- Identification of critical process variables:
    - Detection
    - In-core versus ex-core
        - Profile the core using remote sensing
        - Infer core temperature from pump operations
    - Improved process measurement (reliability, availability, accuracy, etc.)
    - Post-accident monitoring
- Sensors required for online monitoring
- Opportunities for integrated structural sensing
- Test bed design and implementation for sensor development and testing
- Materials research and testing – radiation, temperature, chemistry, etc.
- Uncertainty characterization – signal processing and validation
- Self-calibrating sensors
- Self-testing for sensor failure diagnostics
- Drift-free sensors
- Energy harvesting/scavenging for long-term sensor operation
- In-pile sensors for fuel testing to shorten development time by looking at fuel during irradiation, skipping PIE
- Importing sensors, measurement technologies and data processing from other industries

- Quantum computing integrated into plant control system to secure system signals
- Ultrasound imaging of heat exchanger and reactor vessel during operation (particularly for non-water-cooled reactors)
- RFID sensors for fuel/nuclear material monitoring
- A methodological basis for fusion of probabilistic materials degradation models, data associated with new monitoring technologies, and age-dependent risk models

Finally, researchers must conduct technology readiness assessments of communication needs unique to advanced nuclear applications, with performance requirements. R&D will include tests and evaluations of candidate technologies and architectures in controlled environments and test beds. Standard test protocols for proposed control system and wireless and related communication media will need to also undergo further R&D efforts. R&D for interoperability should include computing systems as well as communications. Researchers need to develop/test/demonstrate:

- Candidate network architectures
- Fault-tolerant network architectures
- Transceiver designs
- Cryptographic, authentication and other secure communication protocols and technologies to ensure data integrity and security
- The electronics needed to transmit signals for on-line monitoring (e.g., modern acoustic emission)

For all sensors and instrumentation systems and associated systems, R&D should address/ include:

- Scoping studies to establish:
    - What needs to be measured
    - Criteria for extended and sensitive advanced systems for *in situ* and real-time monitoring
    - Modeling and simulation needs
- Redundancy versus diversity
- Obsolescence: infrastructure that will accommodate changing technologies and vulnerabilities
- Embedding into design to improve performance and reliability

3. Transformative Nuclear Concepts R&D

Research on transformative nuclear concepts will pursue non-traditional nuclear energy ideas that offer the potential to improve system performance and possibly radically alter nuclear system configuration and development needs. This could include the development of specialized nuclear fuels, revolutionary materials, or coolants; new techniques for energy conversion, waste disposal, or nonproliferation; or other innovations. Pending FY 2011 Congressional appropriation, DOE–NE will support a competitive solicitation process for investigator-initiated projects in the area of transformative concepts.

The purpose of the transformative concepts session was to solicit input on R&D areas needed to transform nuclear energy. The objectives of the panel were to:

- Describe the current and near-term research in reactor and power-conversion technologies, fuels and fuel management, waste disposal, nonproliferation, etc.
- Discuss needs and challenges for nuclear technology that could transform existing or future nuclear systems.
- Obtain input from stakeholders regarding areas for research that could be supported under this program's solicitation.

3.1.1. Overview/Current State of the Art

Reactor and Power Conversion

- The current state of the art is LWR technology in a once-through fuel cycle (open) with a steam plant power conversion system.
- Advanced modeling work is important for robust capabilities, such as lifetime modeling predictions and combined effects. Conservative assumptions are employed for current designs.
- The current state-of-the-art technology in advanced energy conversion is based on conventional steam plant technology (Rankine Cycle). Currently, the focus is on advanced energy conversion technologies for capital cost, safety, efficiency, and compact system benefits beyond the current Rankine cycle. The supercritical CO_2 closed Brayton cycle, in particular, is a promising option (maybe other

supercritical fluids as well). Advanced energy conversion brings safety benefits from fluid compatibility, eliminates any coolant-water reactions, and explores designer fluids for improved heat transfer and compact components.

- Currently, no base technology exists for seismic isolation systems on nuclear plants. Although seismic isolation technologies are established for conventional buildings and structures, they have not been deployed in any current reactor plants within the United States. Seismic isolation may be required for reactors deployed in seismically active areas. Also, SMRs may need seismic isolation systems to be "standardized" and deployable. Containment locations or airplane shields may pose challenges for this technology. Accommodating expansion across the seismic gap may also be a challenge.
- The current state-of-the-art heat exchanger technology is the tube-and-shell heat exchanger. Advanced compact heat exchanger technology can reduce the commodities and size of current base technology. For example, in liquid-metal reactor studies, a hybrid configuration suited for the Na-to-CO_2 heat exchanger is the this heat exchanger with larger formed plate heat exchanger channels on the Na side and smaller printed circuit heat exchangerTM channels on the CO_2 side.

Fuels and Fuel Management

- Within the area of fuels and fuel management, the current goal is development and demonstration of fabrication processes and in-pile performance of advanced fuels/targets (including the cladding) to support the different fuel cycle options (Figure 6) defined in the NE roadmap. Research objectives include development of fuels/targets that:
  - Increase efficiency of nuclear energy production.
  - Maximize utilization of natural resources (uranium, thorium).
  - Minimize generation of high-level nuclear waste (spent fuel).
  - Minimize the risk of nuclear proliferation.

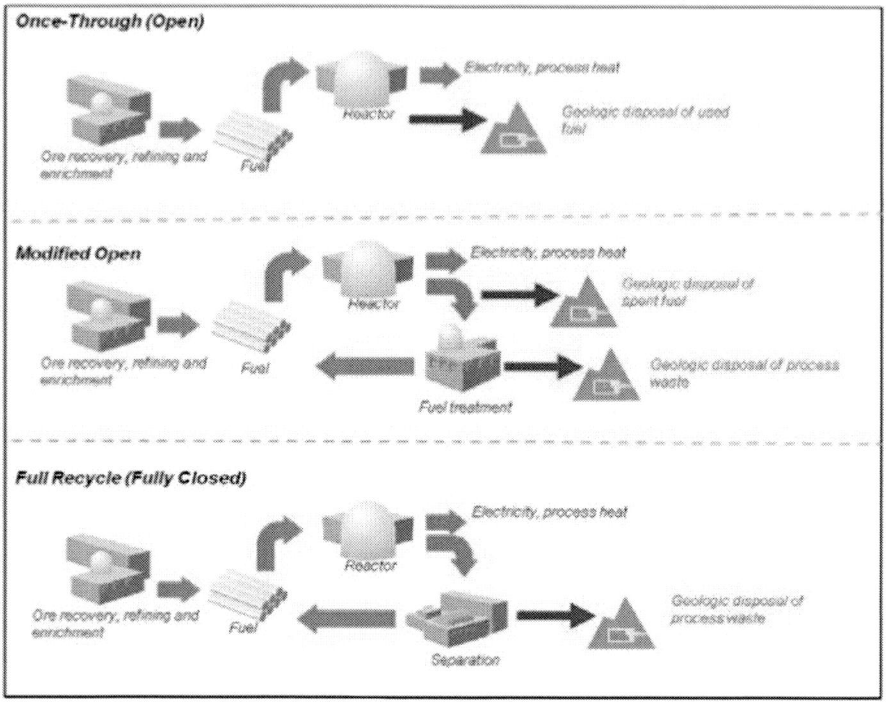

Figure 6. Fuel cycle options.

Waste Disposal

- The mission of the Used Fuel Disposition Campaign is currently to identify alternatives and conduct scientific research and technology development to enable storage, transportation and disposal of used nuclear fuel and wastes generated by existing and future nuclear fuel cycles.
- Interim storage of used nuclear fuel is demonstrated to be safe and licensable. Existing facilities can be licensed to store low-burnup LWR fuel (<45 GWD/MTU) up to 60 years. Additional regulatory and technical bases are needed for interim storage of low-burnup fuels past 60 years and for interim storage of high-burnup fuels.
- Regarding transportation, extensive operational experience exists for LWR fuels and DOE high-level waste. Transportation of high-burnup fuels and advanced fuel forms is less advanced. Secure transportation options are currently limited for Category I and II materials.

- The goals of nuclear waste disposal R&D at this stage are to:
  - Provide confidence that the United States has viable disposal options that will be available when national policy is ready.
  - Identify and research the generic sources of uncertainty that will challenge the viability of disposal concepts and complicate licensing.
  - Increase confidence in the robustness of generic disposal concepts to reduce the impact of unavoidable site-specific complexity.
  - Develop data and a model that will be ready to support repository licensing when needed.

Non-Proliferation

- The current state of the art for non-proliferation includes risk assessment methods, safety and security by design.

3.1.2. Grand Challenges

Reactor and Power Conversion

- Three over-arching grand challenges have been identified in the area of reactor and power conversion:
  - Develop transmutation options that meet a broad range of fuel cycle strategies ranging from deep burn actinide consumption to extended uranium utilization.
  - Develop high-performance transmutation options with usable energy products – comparable to LWR generation costs.
  - Demonstrate prevention of radiation release to public for all events – normal operation, accidents, or malevolent acts.

Fuels and Fuel Management

- The grand challenges for fuels and fuel management are achieving a multi-fold increase in fuel burnup over currently known technologies and achieving a multi-fold decrease in fabrication losses with highly efficient predictable and repeatable processes.
- Within the area of ceramic fuels, grand challenges include:

    - Powder synthesis and optimization for processing and performance testing of feedstock
    - Feedstock/processing sol-gel and resin particle R&D in support of advanced processing and particle fuel development
    - Powder conditioning and process control in processing, which are fundamental to ceramics fuels R&D
- For fuel core materials development, challenges include:
    - A knowledge base up to 200 dpa high-dose core materials irradiation data
    - Development of advanced materials with improved radiation tolerance to greater than 400 dpa under reactor operating conditions

Waste Disposal

- Grand challenges include providing technical bases for confidence in the safety and security of handling spent fuel and wastes from existing and proposed future fuel cycles:
    - Permanent disposal
    - Long-term and extended interim storage
    - Transportation

Non-Proliferation

- R&D must support development of the capability for assured nuclear materials security in future nuclear energy systems through:
    - Next-generation nuclear materials accounting, including more accurate, near real-time, on-line/at-line and nondestructive methods
    - Advanced integrated materials control and process-monitoring technologies, including intelligent data management and analysis, and real-time situational awareness of inventories, operations, activities and anomalies
    - Innovative protection strategies, including transformational breakthroughs in preventing, detecting and responding to attacks

3.1.3. Future R&D

Reactor and Power Conversion

- Future R&D needs include development of:
    - Transformative system (cycle)-level non-traditional nuclear energy concepts that can radically alter nuclear system configuration and development needs and that address the grand challenge of a fully closed fuel cycle.
    - Advanced power conversion technologies to improve upon performance of the Rankine cycle base technology area:
        - One current technology being pursued is the S-CO_2 Brayton cycle technology.
        - This area includes alternative balance-of-plant cooling technologies (minimization of water use).
    - Advanced materials and alloys for structural and heat transfer components:
        - Next-generation materials allow for improved performance, e.g., improved safety margins, longer lifetimes, thinner components, and higher operating temperatures. Example of potential benefits is shown in Table 2.
        - Research needs in development and testing of advanced alloys include creep fatigue testing, environmental effects (coolant, irradiation, temperature), and welding and joining techniques. ASME code qualification of materials is needed for reactor utilization.
- Modeling and simulation research for reactor applications should focus on performance improvements:
    - Higher-fidelity modeling allows for improved performance of existing technology.
    - Improved modeling and simulation facilitates exploration and assessment of innovative configurations and features outside the existing database.
    - Integrated physics modeling in future design tools allows optimization of configuration and performance:
        - Integration of physics, fluids, and structural modeling
        - Tailored systems based on more detailed understanding
        - Streamlined iterations to investigate design refinements

Table 2. Example of advanced materials benefits.

Structure	Parameter	Base	Advanced
Reactor vessel (RV)	Material	316SS	HT-UPS
	Thickness, mm	25.4	15.9
	Mass, kg	255,830	171,914
	Additional savings from IHX	0	-21,322
	Total RV mass, kg	255,830	150,592
Core support structure	Material	316SS	HT-UPS
	Thickness, mm	15.2	10.2
	Mass, kg	109,771	73,483
IHX	Material	304SS	NF616*
	Thickness (tube), mm	1.24	0.889
	IHX mass, kg	52,853	33,730
	Number of IHXs per plant	4	4
	Total IHX mass per plant, kg	211,414	134,920
Intermediate Heat Transport System (IHTS) Piping	Material	316SS	HT-UPS
	Thickness (hot leg/cold leg), mm	25.4/12.7	10.1/9.5
	Mass per loop, kg	53,210	27,344
	Number of loops per plant	4	4
	Total IHTS piping mass, kg	212,842	109,377
Steam generator (SG)	Material	2-1/4Cr-1Mo	NF616*
	Thickness (tube), mm	5.9	2.95
	SG mass, kg	236,009	122,509
	Number of SG's per plant	4	4
	Total SG mass per plant, kg	944,037	490,038
TOTAL MASS OF CONSIDERED STRUCTURES, kg		1,733,894	958,410
	Material savings, kg		775,484
	Material savings, %		44.7%

- Possible R&D for reactor and power conversion includes technology for improving the reliability, maintainability and inspectability of advanced reactor systems and components:
  - Advanced in-service inspection technology for permanent reactor structures
  - *In situ* repair technology for reactor plant systems and components
- Other Grand Challenges and research needs were identified in the DOE's Basic Energy Sciences Advisory Committee workshop held in October 2006.

Fuels and Fuel Management

- Possible R&D for metallic fuels includes:
  - Transformational advancements to achieve high burnup and low process losses:
    - Low-swelling metallic fuels
    - Stable to phase transformations to high burnup
    - Chemical inertness with the fuel cladding
    - Net-shape fabrication
    - Low-temperature fabrication to reduce process losses
  - Modeling and simulation:
    - Atomistic-level understanding of burnup
    - Effects of transmutation elements on phase stability while attaining high burnup
- Possible R&D for ceramic fuels includes:
  - Transformational advancements to achieve high burnup, low process losses, e.g.:
    - Zero-swelling fuels
    - Fission gas capture or release to eliminate fission gas buildup
    - High thermal conductivity fuels
    - Mechanically robust fuels
    - Net-shape fabrication
    - Low-temperature fabrication to reduce process losses
  - Modeling and simulation of ceramic fuels:
    - Understanding high burnup at the atomic scale
    - Effects of transmutation elements on fuel performance
- Within the area of coated particle fuels, needed R&D includes:
  - Transformational advancements to achieve high burnup and low process losses:
    - Robust particle fuels with zero fission gas release to high burnup
    - Additives to getter fission gases during burnup
    - Processes for uniform-coated-particle production
  - Modeling and simulation:
    - Modeling burnup in coated particles at the atomic scale
    - Predicting particle fuel failure at high burnup
- Within the area of irradiation and testing of advanced fuels, needed R&D includes:

- Transformational advancements to irradiate to high burnup, including:
    - Developing small-scale techniques to gain a fundamental understanding of high burnup in advanced fuels
    - Developing techniques for accelerating achievement of high burnup on advanced fuels
    - *In situ* physical measurements under irradiation down to atomistic scale
- Transformational advancements to test fuel property changes after high burnup, such as:
    - Developing techniques for quick and accurate measurement of physical properties on irradiated fuels
    - Developing small-scale techniques for testing physical properties on fuels after high burnup irradiation

Waste Disposal

- R&D needed for used nuclear fuel storage includes:
    - Degradation and aging phenomena, such as:
        - Aging of fuel, cladding, and primary container
        - Aging of storage site infrastructure
        - Long-term monitoring and NDE techniques
    - Radiological consequences of storage, such as:
        - Advanced modeling for aircraft crash analysis
        - Advanced modeling for sabotage event analysis
        - Integration of security requirements with storage design and operations
        - Advanced modeling for dispersion of radioactive materials
- R&D needed for used nuclear fuel transportation includes:
    - Risk-informed cask qualification, including modeling and data
    - Radiological consequences of storage, including:
        - Advanced modeling for dispersion of radioactive materials following accident or intentional destructive acts
        - Data for behavior of high-burnup and advanced fuels
- R&D needed for used nuclear fuel disposal includes:
        - Waste form performance, e.g., volume, activity, thermal output, durability in a range of environmental conditions

- Container materials for alternative disposal concepts; evaluation of material performance in a range of environmental conditions
- Advanced modeling and simulation tools to evaluate performance of generic disposal concepts
- Field and laboratory experiments to validate models for generic disposal concepts
- Operational design validation for alternative disposal concepts, e.g., deep borehole disposal concepts

Non-Proliferation

- With regard to MPACT, future R&D should include development of laser ablation methodology to determine the elemental composition and plutonium/uranium content in hulls. Laser-induced breakdown spectroscopy (LIBS) can use atomic emission, Raman, or structural characterization of a solid sample.
- MPACT should integrate visible and Raman spectroscopic techniques with macroscopic solution property measurements (non-invasive ultrasound-based viscosity, density, conductivity, temperature, refractive index) in an on-line process monitor for spent nuclear fuel safeguards. The research community should focus on:
  - Preparing a database of spectral and physical property measurements (density, conductivity, temperature, etc.) over a wide range of plutonium and minor actinide solution compositions.
  - Developing interpretive and predictive models for the quantification of plutonium and minor actinides based on multiple physical property measurements and selected inputs from spectroscopic sensors.
  - Validating this new technology against current methods.
  - Translating the laboratory system into online real-time configuration.
- On-site mini-processing of nuclear waste (one spent rod at a time) to separate and vitrify wastes using modern directed energy tools would provide a politically acceptable solution to nuclear waste disposal.

4. Summary and Conclusions

The NEET program proposed for FY 2011 will support development of crosscutting technologies that directly support and complement DOE–NE's advanced reactor and fuel cycle concepts, focusing on innovative research that offers the promise of dramatically improved performance.

The workshop successfully promoted discussion on what research will be needed under the new NEET program. This workshop was not designed to solve any of the problems facing the nuclear industry, nor was the goal to produce a final list of research topics. Rather, the workshop was intended to encourage stakeholder participation and gather ideas for research that is needed to advance nuclear technology in support of multiple reactors and fuel cycle concepts.

Based on the panel sessions, below is a summary of the relevant findings in each technical area.

Innovations in manufacturing and construction have led to the ability to construct new nuclear power plants more efficiently and reliably than ever before. The future of the industry is modular design, along with advances in composite structures, seismic base isolation, and pre-assembled rebar systems. Innovations in transportation, lifting and rigging, materials, welding, and modeling will also have profound impacts. However, new products and processes are irrelevant if they do not receive regulatory approval; technology must first pass through the regulatory process for out-of-the-box ideas to be considered.

In addition, R&D must take a holistic approach. For example, construction code cases that replace radiographic inspection with ultrasonic techniques offer secondary implications for online monitoring. Alternate monitoring capabilities, in turn, could affect pathways for theft of materials. Virtual environments and other integrated programs could help examine multiple aspects of next-generation technologies more holistically.

Minimizing the risks of nuclear proliferation and terrorism is an integral part of NE's RD&D mission and program. Science-based investigator-driven R&D will focus on key challenges: for example, adversary decision models for proliferation and terrorism, mathematical models for performing risk analysis, and treatment of uncertainty. Additionally, focus on expert elicitation techniques in the proliferation/terrorism context and effective communication of results will be an important part of these efforts. In parallel, risk assessment tools should be applied to high- priority problems and issues to help demonstrate the value and uses of such tools and methodologies while highlighting gaps and informing longer-term R&D. The program will utilize cross-disciplinary teams (including political scientists and social scientists) and solicit a broad spectrum of stakeholder input including expertise from, and benchmarking against, risk management techniques used in other fields. Ultimately, these R&D efforts will lead to innovative, vital, and indispensable tools and methodologies for assessing, comparing, and managing the proliferation and terrorism risks of nuclear energy technology, fuel cycle systems, and global nuclear energy architecture options.

Further research is required to both enhance current materials and develop or uncover new materials, such as advanced composites. Composites have the potential to offer designers substantially different properties from conventional homogeneous materials, altering such basic characteristics as ductility and strength. The challenge in the nuclear industry will be convincing designers to consider composites in new applications.

Although every material will have different testing requirements, not every material will need to receive the same battery of tests. Proposals should outline the possibilities for each material, as well as what testing has already been done and what remains to be done. New characterization tools and new modeling and simulation tools are cross-cutting, but performance requirements do not necessarily overlap different reactor types. Since materials will have specific applications, separating program goals from the goals for specific materials will be difficult. In a high-risk– high-reward environment, R&D could generate materials with cross-cutting applications.

Needs for advancement in instrumentation, controls, and related technologies were identified across all of DOE's nuclear energy programs. These include needs for new sensors to accurately measure nuclear process behavior as well as reduce the uncertainty in measurements obtained. These sensors will be able to withstand high temperatures and radiation and measure system properties directly. Sufficient cross compatibility does not exist between sensors in use today for nuclear process measurement and those that will be needed for the kinds of processes and environmental conditions envisioned by nuclear energy programs.

In new nuclear power plants, monitoring and control technologies will be based on different technologies and act differently than their counterparts in existing nuclear power plants. New systems will be digital and incorporate automation to achieve high reliability, safety, and enhance plant efficiency. The focus of new monitoring and control technologies will be on resilience; they must be able to reliably perform in high noise conditions, offer greater functionality, be mode aware, and function collectively using intelligent technologies. Future nuclear energy systems should also integrate systems and functions that are today independent and isolated. This includes multiple operating nuclear units, applications of nuclear heat for new processes (e.g., industrial heat, desalination, etc.) and electricity. Needs for integrated control, communications, data transmission for digital technologies, and their qualification for nuclear deployment can best be addressed through a program of crosscutting R&D.

These new technologies must be considered for implementation at the design stage. Modular components, digital instrumentation and controls, and PR&PP features should not be "paste-ons" but incorporated at the earliest stages of design. Reactor materials research needs to be "ahead of the curve" so that vendors can take new materials properties into consideration while developing new designs and allowing time for NRC certification.

The transformative nuclear concepts R&D session goal was to identify R&D areas in which non-traditional nuclear energy concepts could offer the potential to radically alter nuclear system configuration and development needs. To help determine what projects will ultimately move forward, carefully developed metrics and criteria were recommended as tools to rank and select transformative projects. The research areas presented in this session were only a representation of grand challenges. Numerous other high-risk–high-reward concepts have the potential for enabling significant leaps forward in advanced nuclear technology development. These concepts will be the focus of the open solicitation pending FY 2011 appropriation.

All ideas gathered during this workshop will help NE shape the new NEET program for FY 2011. Collaborative efforts must continue to ensure appropriate crosscutting research and technology development relevant to the various reactor and fuel cycle concepts within the scope of NE R&D programs. The activities undertaken in this program must complement those within the Reactor Concepts RD&D and Fuel Cycle R&D programs by providing a mechanism for pursuing broadly applicable R&D in areas that may ultimately benefit specific reactor and nuclear fuel concepts. Leveraging the knowledge generated through activities in the NEET program will provide useful information for program and strategic planning and will allow NE to address key challenges affecting nuclear reactor deployment (e.g., capital cost, technology risks, and proliferation concerns).

APPENDIX 1 – BREAKOUT SESSIONS

U.S. Department of Energy, Office of Nuclear Energy (DOE–NE) Nuclear Energy Enabling Technologies (NEET) Program Workshop

Advanced Methods for Manufacturing and Construction

Moderator: Mr. Jack Lance, Idaho National Laboratory

Panel: Mr. Nate Ames, Edison Welding Institute (EWI), Nuclear Fabrication Consortium
Mr. Kenneth Barry, Electric Power Research Institute (EPRI), Advanced Nuclear Technology Program
Mr. Craig Hanson, Babcock & Wilcox (B&W)
Mr. John Simmons, URS Washington Group

Goals and Objectives

This session will explore areas where new and innovative technologies can be developed to support the continuing growth of the U.S. nuclear power industry. Focus is on the areas of advanced manufacturing and fabrication technologies and advanced construction methods.

The goal is to reduce cost and schedule for new nuclear plant construction and make fabrication of NPP components faster and cheaper with better reliability. Ultimately the goal is to restore the U.S. position as a manufacturer and constructor of NPP designs in the U.S. and the world. Fabrication innovation can include welding technologies that use multiple processes including lasers, integrated welding and NDE systems, advances in pipe fabrication and welding to support modular build, advances in large machining processes such as ultrasonic enhanced machining, wiring and cabling systems that support modular builds, fast and reliable material characterization systems that allow for material identification at receipt and on the job, in seconds with documentation, and innovations in pipe and component insulation systems.

Construction innovation will take modular building to a higher level with advances in composite structures, base isolation systems for seismic integrity, rebar systems that are pre-engineered and assembled, possibly using new rebar materials; innovations in transport, lifting and rigging activities; innovation in structural materials and how they are used; new practices in field assembly and welding and in modeling to improve the design, construction, and project management of large-scale construction.

Proliferation Risk Assessment

Moderator: Dr. Sara Scott, Los Alamos National Laboratory
Panel: Dr. Robert Bari, Brookhaven National Laboratory
Dr. William Charlton, Texas A&M University
Dr. William Burchill, Past President, American Nuclear Society Mr. Mark Whitney, National Nuclear Security Administration

Goals and Objectives

This session will solicit views of a broad cross-section of stakeholders on the following questions:

- Would you favor an expanded R&D effort on proliferation and terrorism risk assessment? Why or why not?
- In what ways have current Proliferation and Terrorism Risk Assessment Methodologies been useful, and how might research funding make them more effective?
- If an expanded R&D program were initiated, what aspects of proliferation and terrorism risk assessment should receive priority attention? What are the most promising or fruitful areas for such research? What are the least promising or least worthwhile? If a mix of topics were considered, what would be appropriate elements of a balanced portfolio?
- If an expanded R&D program were initiated, what cautions and recommendations should DOE–NE bear in mind as they plan and implement the program?

The session will identify promising R&D opportunities for a new initiative on proliferation/terrorism risk assessment R&D.

Reactor Materials

Moderator: Ms. Sue Lesica, U.S. Department of Energy
Panel: Dr. Jeremy Busby, Oak Ridge National Laboratory
Dr. Michael Burke, Westinghouse

Dr. Stuart Maloy, Los Alamos National Laboratory
Dr. Arthur Motta, Penn State University

Goals and Objectives

Develop new classes of alloys and materials, not yet considered for reactor performance that may enable transformational reactor performance. The custom design of innovative steels using modern materials science techniques, industrial knowledge, and previous experience can improve performance over traditional materials by a factor of five to ten, increasing the maximum operating temperature by 200 degrees Celsius for a period of at least 80 years. Concepts that may be evaluated include optimized alloy composition, engineered microstructures, age- tempered microstructures, or combinations thereof. Other, more radical concepts that may be explored to enable even greater performance include bimetallic layers, metal/ceramic composites, ion-beam or surface-modified alloys. A wide range of operating conditions will be considered, with the general goal of improved strength and radiation resistance.

Objectives are as follows:

- Provide a description of the proposed scope of activities in this crosscutting program related to innovative reactor materials and their relevance to DOE sponsored programs.
- Delineate between the technical activities proposed in this program element and program elements related to innovative reactor materials in other DOE sponsored R&D programs.
- Discuss and receive input on gaps between needed and existing capabilities that may be addressed through this R&D program.
- Identify areas for coordinated research between this proposed R&D program and other planned or ongoing research and development efforts.

Advanced Sensors and Instrumentation

Moderator: Dr. Bruce Hallbert, Idaho National Laboratory
Panel: Dr. Richard Wood, Oak Ridge National Laboratory
Dr. Joseph Naser, Electric Power Research Institute
Dr. Wes Hines, University of Tennessee
Dr. John Collins, Idaho National Laboratory

Goals and Objectives

Inform stakeholders about the status of planning for a multi-year program of research and development on sensors, instrumentation, and related technologies that crosscut DOE nuclear energy programs. Solicit input on out-year plans for research and development needed to achieve end-state capabilities that can be applied to nuclear energy development programs.

Objectives are as follows:

- Provide a description of the proposed scope of activities in this crosscutting program related to sensors, instrumentation, and related technologies and their relevance to DOE sponsored programs.
- Delineate between the technical activities proposed in this program element and program elements related to instrumentation, controls, and related technologies in other DOE-sponsored R&D programs.
- Discuss and receive input on gaps between needed and existing capabilities that may be addressed through this R&D program.
- Identify areas for coordinated research between this proposed R&D program and other planned or ongoing research and development efforts.
- Obtain input from stakeholders regarding (1) existing capabilities and programs related to sensors and instrumentation; (2) private sector interest in opportunities for cost-shared R&D; (3) university interest in educational programs that may be sponsored under this program.

Transformative Nuclear Concepts Research and Development Panel

Moderator: Dr. Kemal Pasamehmetoglu, Idaho National Laboratory

Panel: 1. Reactor and Power Conversion: Mr. Christopher Grandy, Argonne National Laboratory
2.Fuels and Fuel Management: Dr. Stuart Maloy, Los Alamos National Laboratory
3.Waste Disposal: Dr. Peter Swift, Sandia National Laboratories
4.Non-Proliferation: Mr. Mark Mullen, Los Alamos National Laboratory

Goals and Objectives

The goals of this panel are as follows:

- Inform stakeholders on the proposed open competitive solicitation, pending FY 2011 Appropriation, for projects that relate to any aspect of nuclear energy generation — reactor and power conversion technologies, fuels and fuel management, waste disposal, nonproliferation, and so forth — ensuring that good ideas have sufficient outlet for exploration.
- Solicit input on research and development areas/gaps needed to transform nuclear energy. The research on transformative nuclear concepts will pursue non-traditional nuclear energy ideas that offer the potential for improved system performance and may radically alter nuclear system configuration and development needs. This could include the development of specialized nuclear fuels, revolutionary materials, tailored coolants, new techniques for energy conversion, or other innovations.

The objectives of this panel are as follows:

- Provide a description of the current and near-term research in reactor and power conversion technologies, fuels and fuel management, waste disposal, nonproliferation, among others.
- Discuss gaps and needs for nuclear technology that could transform existing or future nuclear systems.
- Obtain input from stakeholders regarding areas for research that could be supported under this program's solicitation.

INDEX

D

E

F

G

H

I

Q

R

S

T

U